AF545066

Contemporary Science of Polymeric Materials

A Symposium in honor of Professor Frank E. Karasz on the occasion of his 75th birthday, Valletta, Malta, February 28–March 2, 2009

ACS SYMPOSIUM SERIES **1061**

Contemporary Science of Polymeric Materials

A Symposium in honor of Professor Frank E. Karasz on the occasion of his 75th birthday, Valletta, Malta, February 28–March 2, 2009

Ljiljana Korugic-Karasz, Editor

Sponsored by the ACS Division of Polymer Chemistry

American Chemical Society, Washington, DC

Library of Congress Cataloging-in-Publication Data

Contemporary science of polymeric materials : a symposium in honor of Professor Frank E. Arasz on the occasion of his 75th birthday, Valletta, Malta, February 28-March 2, 2009 / Ljiljana Korugic-Karasz, editor ; sponsored by the ACS Division of Polymer Chemistry.

p. cm. -- (ACS symposium series ; 1061)

Includes bibliographical references and index.

ISBN 978-0-8412-2602-9 (alk. paper)

1. Polymers--Congresses. 2. Polymeric composites--Congresses. I. Korugic-Karasz, Ljiljana S., 1953- II. Karasz, Frank E., 1933- III. American Chemical Society. Division of Polymer Chemistry.

TA455.P58C67 2010

620.1′92--dc22

2010050325

The paper used in this publication meets the minimum requirements of American National Standard for Information Sciences—Permanence of Paper for Printed Library Materials, ANSI Z39.48n1984.

Distributed by Oxford University Press

PRINTED IN THE UNITED STATES OF AMERICA

Foreword

The ACS Symposium Series was first published in 1974 to provide a mechanism for publishing symposia quickly in book form. The purpose of the series is to publish timely, comprehensive books developed from the ACS sponsored symposia based on current scientific research. Occasionally, books are developed from symposia sponsored by other organizations when the topic is of keen interest to the chemistry audience.

Before agreeing to publish a book, the proposed table of contents is reviewed for appropriate and comprehensive coverage and for interest to the audience. Some papers may be excluded to better focus the book; others may be added to provide comprehensiveness. When appropriate, overview or introductory chapters are added. Drafts of chapters are peer-reviewed prior to final acceptance or rejection, and manuscripts are prepared in camera-ready format.

As a rule, only original research papers and original review papers are included in the volumes. Verbatim reproductions of previous published papers are not accepted.

ACS Books Department

Contents

Self-Assembly and Polymer Networks

Simulation and Modeling in Polymer Science

Indexes

Preface

The chapters in this volume are representative of the contributions presented at a symposium, “Contemporary Science of Polymeric Materials” held in Valletta, Malta, February 28 – March 2, 2009 on the occasion of 75th birthday of Frank E. Karasz.

Frank (Franz) Erwin Karasz was born in Vienna. His undergraduate education was at the Royal College of Science, Imperial College, University of London, where he graduated with honors in Chemistry. His Ph.D. thesis at the University of Washington was concerned with the statistical mechanics of liquid argon/helium mixtures. After a year’s post-doctoral work with the well known biopolymer theoretician Terrell Hill at the University of Oregon, he returned to England to the National Physical Laboratory, near London. There he worked on the thermodynamics of folded chain single crystal polyethylene and started a collaboration with the late John Pople which resulted in a number of theoretical papers dealing with transitions in rotationally disordered (“liquid”) crystals. Remarkably the basic Karasz-Pople theory is still somewhat in vogue after more than 50 years with commentaries and modifications appearing every now and then.

After two and a half years at the NPL, Karasz returned to the U.S. to take a position at the General Electric Research Laboratory in Schenectady, New York. One of his interests in that era was in the helix-coil or order-disorder transition in synthetic polypeptides. He and his collaborator, James O’Reilly, were in fact the first to directly measure the transition enthalpy in such a system, solubilized poly-γ-benzyl-L-glutamate. He also became interested in polymer blends, which were beginning to be of scientific and theoretical interest.

Karasz left G.E. in 1967 and went to the University of Massachusetts where he and many collaborators collaborated in an extensive set of studies of polymer blends and alloys. A major advance in this area was made in 1982 when, with William MacKnight and Gerrit ten Brinke, a comprehensive quantitative theory of copolymer blends was developed which had both interpretive and predictive value. Their mean field theory was successfully applied to a wide variety of systems, and has been used by many others since then.

In 1972, Karasz and the late Roger Porter established an NSF Materials Research Laboratory at the University of Massachusetts. As a result, around 1979, he became acquainted with the work of Alan Heeger and Alan MacDiarmid, both at that time at the University of Pennsylvania. This, together with colleague James Chien led to an extensive collaboration in the rapidly developing field of π-conjugated conducting polymers especially polyacetylene. Karasz and his co-workers continued to carry out research in this area and extended research to the study of electroluminescence of poly phenylene vinylene and other conjugated polymers. In 1993 Karasz produced the first electroluminescent polymer to emit blue light and he continued, again with a long list of co-workers, to be a major contributor to this field.

In 2004, Karasz formally retired from the Silvio O. Conte Chair in the Department of Polymer Science and Engineering at the University of Massachusetts. However, he continued an extensive research program in the optical properties of π-conjugated polymers and currently is interested in the thermo-electric properties of these materials. Activity in inorganic thermoelectric materials research has increased because of wide potential applications in energy conversion and related areas. Thus the development of a viable functional polymeric thermoelectric material will have very broad implications. His program on TE polymers is now expanding with an emphasis on hybrid and blended macromolecular materials, which have the advantages of an inherently low thermal conductivity and a relatively facile processability.

Karasz's first scientific paper was on the adsorption of water on AgI and TiO_2, published in 1956; thus he is now in his sixth decade of a prolific publishing career, with a total of about 550 papers to date. He has received substantial recognition starting with the Mettler Medal in 1975. In 1984 he and W.J. MacKnight were co-recipients of the Ford Prize of the American Physical Society, and in 1985 he received the Research Award from the Society of Plastic Engineers. He has received numerous other awards including the University of Ferrara 600th Year Commemorative Medal, and in 2000 Polytechnic University presented him the SEAM Award. In 2002 he received the Herman Mark Medal from the Austrian Institute for Chemical Research.

In 1991 Karasz was elected to the U.S. National Academy of Engineering and since then he has been elected to membership in the Croatian, Indian and Serbian Academies of Science and Art. He is also a Chevalier of the Knights of St. John and a Fellow of the American Physical Society, the AAAS, and the PMSE Division of the ACS.

Karasz has long had a strong relationship with the European polymer community and for some time served, for example, as President of the Scientific Council of the CNR Polymer Laboratory in Naples. He is on the Editorial Boards of several European journals and continues to be very actively involved at the frontiers of polymer research.

Ljiljana Korugic-Karasz

Polymers for Environmentally Sustainable Applications

Chapter 1

Sugar-Based Chemicals for Environmentally Sustainable Applications

Xianhong Feng, Anthony J. East, Willis Hammond, and Michael Jaffe*

**Medical Device Concept Laboratory,
Department of Biomedical Engineering, New Jersey Institute of Technology,
111 Lock Street, Newark, NJ 07103**
***jaffe@adm.njit.edu**

Sugar-based chemicals can be used as building blocks to produce new monomers, polymers and additives for the commercial plastics and cosmetics industry. As GRAS materials, i.e, "generally recognized as safe", sugar compounds with a unique stereochemistry provide a ubiquitous platform for making cost effective chemicals and polymers. Isosorbide and its isomers as sugar derived dianhydrohexitols can be either incorporated the backbone of new polymers or converted to low molar mass additives for thermoplastics and thermosets or as specialty chemicals. As the cost of petroleum rises, the attractiveness of renewable feedstocks for producing value added products increases. The emergence of sustainable sugar derived chemicals (especially isosorbide modified products) offers attractive prospects with high potential for the next generation chemical industry.

Introduction

If the 20th century was the era of petroleum refining, then the 21st century may be the era of the "bio-refining" industry (*1*, *2*). Before it was recognized that petroleum is not an inexhaustible and universal source of energy, the petroleum industry was growing dramatically with chemical synthesis of new products with novel properties, particularly polymers and plastics (*3*, *4*). Since World War II, the petroleum based polymer industry has benefited modern society in numerous ways in our everyday environment (*5*). From packaging to construction materials, the specially designed polymers with the property of disposability provided new possibilities for replacing conventional material like glass and metals and bringing about advances in convenience. In the 1970s, the first oil crisis reoriented the energy and chemical industries. The increasing demand for limited petroleum resources accounted for the rising price of fossil fuels (*6*, *7*). The environmental contamination from discarded polymeric products at the end of their useful life drove the "green" movement and encouraged the chemical industry towards sustainable development (*8–12*). The multiple concerns of economy, environmental and societal responsibility lead to a humanitarian concept aimed at the improvement in quality of life and welfare of humanity without sacrificing the needs of future generations (*13*). There was a growing urgency to seek the renewable sources and convert them to useful products including chemicals, energy sources and materials for replacing the enormous demand for fossil fuels (*14*). Agricultural feedstocks as natural, renewable and sustainable resources have potential for use as biopolymers, bioenergy sources and bio-chemicals (*15–18*). The primary production of biomass estimated in energy equivalents is 6.9×10^{17} kcal/year over the entire world. Only 7% of the energy was utilized by human beings. In the United States, sugar-based biomass generated from corn, soybean, wheat and sorghum, the four major crops grown in the country (*19*), was faced with the dual challenges of oversupply and low prices (*20*). An integrated approach was needed for new high value products, new uses and new concepts in sugar manufacturing (*21*). In order to circumvent difficulties in controlling the reactivity of the many different hydroxyl groups on carbohydrates, simple molecules such as 1,4:3,6-dianhydrohexitols with two hydroxyl groups have explored as suitable starting materials for new monomers, polymers and additives for the commercial plastics and cosmetics industry.

Dianhydrohexitols are dihydroxyethers made by dehydrating hexitols which are polyhydric alcohols derived from hexose sugars by reduction, chiefly glucose, mannose and idose (*22–25*). Since each of the dianhydrohexitols has two free hydroxyl groups located at the C-2 and C-5 positions either inside (named as ***endo***) or outside (named as ***exo***) the bicyclic ether ring (*26*), its three diastereoisomers known as 1,4:3,6-Dianhydro-D-glucitol (isosorbide **1**), 1,4:3,6-Dianhydro-D-mannitol (isomannide **2**) and 1,4:3,6-Dianhydro-L-iditol (isoidide **3**) can be used as different monomers when they are incorporated into polymers as diols. (Figure 1). The hydrogen bonding between the ***endo***-hydroxyl group and the ether oxygen makes the ***exo***-hydroxyl have different reactivity to that of the ***endo***- hydroxy group (*27*, *28*). Considering the steric effects and hydrogen bonding, isomannide with two ***endo*** hydroxyl groups is the least reactive compound compared with isoidide which has two ***exo*** hydroxyl groups and should be the more attractive in biochemical applications (*29*). Unfortunately, isoidide is rare in nature and still remains expensive in many applications especially as a monomer in the polymer industry (*30*). For isosorbide, the most widely available dianhydrohexitol, the hydroxyl group at C-2 is ***exo*** and that at C-5 is ***endo*** (*31*). Since it is classified by the Food and Drug Administration as a "general recognized as safe", GRAS, material and can be made readily available, isosorbide has potential for use as a "green" alternatives to petroleum based chemicals and polymers (*32*). Its molecular geometry and chemical functionality provides its compatibility with many commercial plastics and specialty additives. The asymmetric reactivity, chirality and controlled stereochemistry in the design and performance of its derivatives, including thermoplastics, thermosets and low molar mass compounds that can act as plasticizers, stabilizers or compatiblizers, make it more commercially attractive in the polymer and specialty chemical industry.

Isosorbide Based Thermoset

There is a growing awareness that Bisphenol-A (BPA) is a xenoestrogen, a molecule that possesses the ability to mimic estrogen and cause feminizing effects in living creatures, particularly male animals including humans (*33*, *34*). BPA is widely used for coatings in food packaging and a key ingredient in plastics ranging from dental composites to baby bottles. Since the chemical bond between the BPA and its linked polymer is not completely stable, the polymer could slowly decay with time and release small amounts of BPA which can contaminate materials with which it comes into contact, such as food or water (*35*, *36*). There are over six-billion pounds of BPA containing materials are manufactured every year. The spread of a tiny amount of BPA in the environment could exert its estrogen-like effects leading to extensive human and animal exposure. The growing environmental consciousness necessitates the bio-derived replacement of BPA in the near future. The major opportunity here for isosorbide is its remarkable chemical properties and attractive price. As a biodegradable and naturally derived material, isosorbide is a rigid organic diol with the similar structure to that of BPA, but without the endocrine disrupting effect. Isosorbide can be attached to glycidyl ether or allyl ether to make crosslinkable epoxy resin monomer with similar properties to bis-A diglycidyl ether.

The monoisosorbide diglycidyl ether was prepared by a Williamson ether reaction. Diallyl isosorbide ether was prepared by heating the isosorbide with allyl bromide in sodium hydroxide solution as shown in Scheme1. Freshly-prepared unpurified diallyl isosorbide was treated with the *meta*-chloroperbenzoic acid in methylene chloride to generate isosorbide diglycidyl ether as shown in Scheme 2. Since the isosorbide has two hydroxyl groups with different reactivity, by using the aqueous alkali, the ***exo***-hydroxyl group of isosorbide could be alkylated first leaving the ***endo***-hydroxyl group of two isosorbide molecules being linked with one epoxide as shown in Scheme 3. The bisisosorbide diglycidyl ether was prepared by heating the isosorbide with 50% sodium hydroxide solution and a large excess of epichlorhydrin, which was used to azeotrope away the water. Two equivalents isosorbide are linked by three molecules of epichlohydrin to form the epoxide dimer.

The epoxy equivalent of the monomer was measured according to ASTM D1652-04 by reaction with an aliquot of standard pyridine hydrochloride in excess pyridine at 100℃ under nitrogen and subsequent back titration with standard methanolic potassium hydroxide using phenolphthalein as indicator.

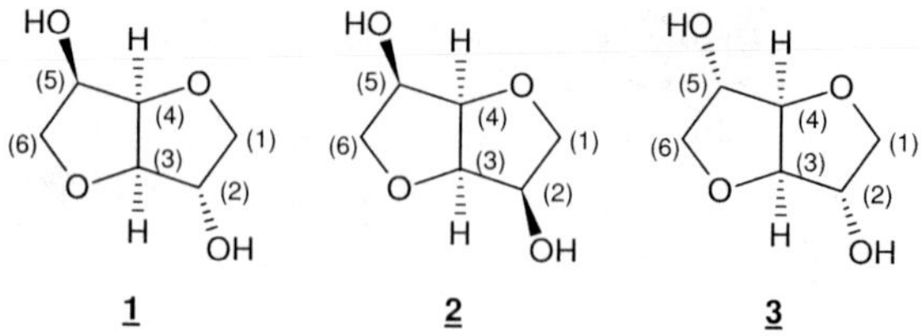

1 **2** **3**

Figure 1. The three diastereoisomers of dianhydrohexitol

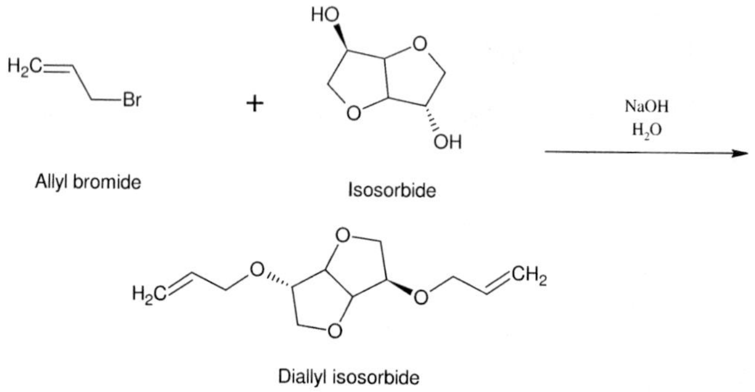

Allyl bromide + Isosorbide $\xrightarrow{NaOH, H_2O}$ Diallyl isosorbide

Scheme 1. Synthetic route of diallyl isosorbide

Step ②:

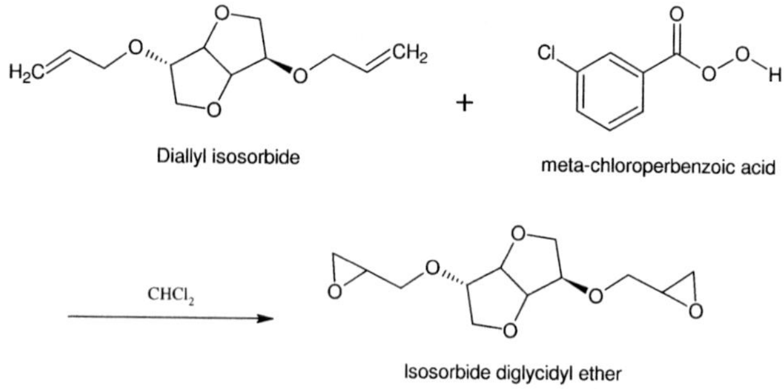

Diallyl isosorbide + meta-chloroperbenzoic acid $\xrightarrow{CHCl_2}$ Isosorbide diglycidyl ether

Scheme 2. Synthetic route of isosorbide diglycidyl ether

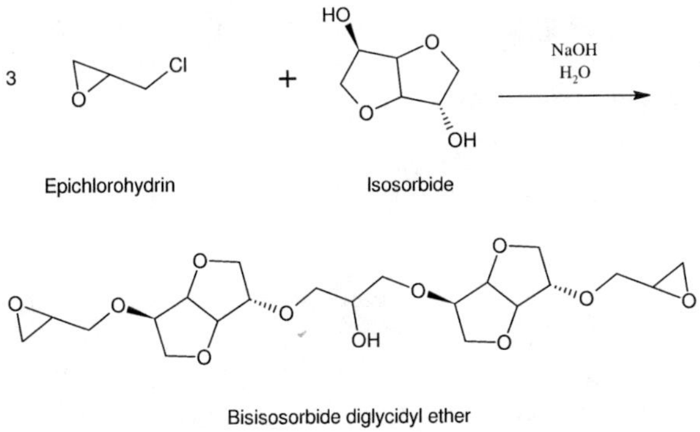

Epichlorohydrin Isosorbide

Bisisosorbide diglycidyl ether

Scheme 3. Synthetic route of bisisosorbide diglycidyl ether

The monoisosorbide diglycidyl ether with the epoxide equivalent weight of 129 Daltons indicates the two hydroxyl groups of isosorbide are linked with two epoxides. The bisisosorbide diglycidyl ether with epoxide equivalent weight of 223 Daltons corresponding to 446 Daltons of molecular weight of monomer, indicates a dimeric structure with two isosorbide units joined with a 2-hydroxy-1,3-propane diether link and capped with glycidyl ether units.

With different hardening agents, the pure isosorbide epoxy resin can be converted from the thermoplastic state to tough, hard, thermosets. Using the primary and secondary amines or anhydrides as hardeners for different crosslinking applications, a broad range of isosorbide derived epoxys can be made with properties such as excellent adhesion, high chemical or heat resistance and good-to-excellent mechanical properties.

When bisisosorbide diglycidyl ether was cured with an aliphatic amine, Jeffamine T403, a "green" commercial liquid curing agent, the tensile strength of the resulting isosorbide epoxy is 104% of the thermoset obtained from EPON 826, a diglycidyl ether of bisphenol A (DGEBA) type commercial epoxy resin. The impact strength of isosorbide epoxy is 40% higher than the Bis-A epoxy. The good mechanical properties of Jeffamine cured isosorbide epoxy provides a new class of environmentally friendly thermosets, which could be used as coatings, adhesives and composites. However, its glass transition temperature, measured by DSC as the typical step change in the heat flow curve, was shown to be 48℃, which is much lower than DGEBA epoxy of 90℃. The depressed Tg relates to isosorbide's high affinity for water. When different crosslinkers were used to cure isosorbide epoxy resin, the more hydrophobic crosslinking agents like 4,4′-(hexafluoro-isopropylidene) diphthalic anhydride with higher crosslinking density were sufficient to offset the hydrophilic property of isosorbide glycidyl ether to raise its Tg to 200℃. The water uptake ratio is believed to be determined by the factors of crosslinking density, the chemistry of crosslinking agent and the amount of free hydroxyl groups on the backbone of the epoxy. By adding the hydrophobic functional group into the backbone of isosorbide epoxy or adjusting the amount and type of crosslinker, the mechanical property and water uptake ratio of the isosorbide derived epoxy could be optimized for different applications. The high water uptake epoxy with controllable biodegradation rate could be used as a drug delivery system or extracellular matrix for biomedical applications while the low water uptake epoxy with strong mechanical properties could be used for can coatings, bone cements and other industrial additives and adhesives.

Isosorbide Based Humectant

Humectants are widely used in numerous industries, e.g. food additives, cosmetic moisturizers and pharmaceutical formulations. Their highly hygroscopic properties mean that they are able to absorb water from the atmosphere. Frequently they are molecules with several hydrophilic groups, most often hydroxyl groups and carboxyl groups. These provide an affinity for forming hydrogen bonds with molecules of water. As described before, isosorbide diglycidyl ether is a very hydroscopic material. Its major component, isosorbide, is also known to be very hygroscopic and good candidate for a humectant. One liter of water can dissolve 8 kg of isosorbide which functions like a cyclized polyethylene glycol (PEG) (*37*). When isosorbide is chemically attached to a glycidyl ether, not only can a range of epoxides be prepared, but also a series of polyols can be generated by hydrolyzing their corresponding epoxides with dilute acid. The chemical structure of these hydrolyzed polyols is similar to a polymeric glycerol when isosorbide is attached with the glycerine (propanetriol), a known skin softener and humectant. The hygroscopic properties of such polyols would be expected to be boosted tremendously and a new class of highly effective humectants as "green" replacements for PEG could be prepared for cosmetic formulations. Whereas PEG is derived from non-renewable petroleum sources, isosorbide based polyols are derived from renewable sugars.

As shown in Scheme 4 and 5, isosorbide derived polyols were prepared by ring-opening the epoxides of isosorbide glycidyl ethers with water containing a trace of trifluoroacetic acid as a catalyst. The product, bisisosorbide triglycerol, was obtained as yellow liquid with a higher viscosity than the isosorbide diglycerol, a colorless product.

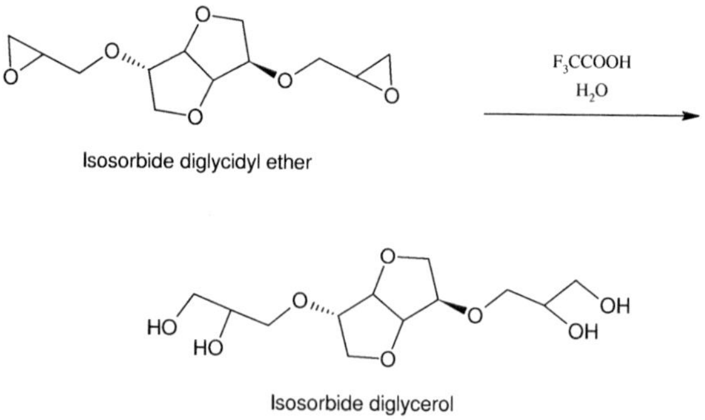

Isosorbide diglycidyl ether

F_3CCOOH
H_2O

Isosorbide diglycerol

Scheme 4. Synthetic route of isosorbide diglycerol

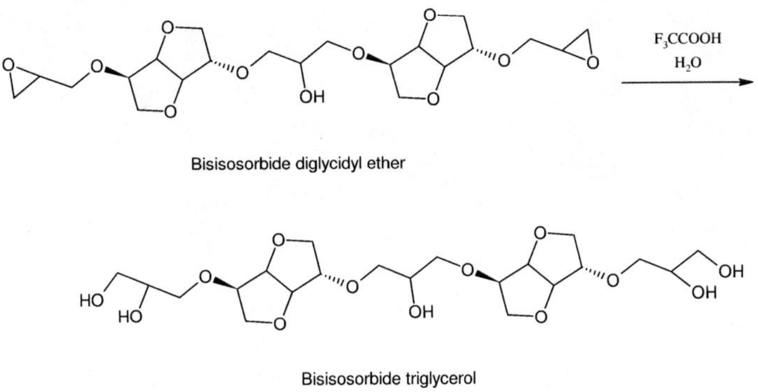

F_3CCOOH
H_2O

Bisisosorbide diglycidyl ether

Bisisosorbide triglycerol

Scheme 5. Synthetic route of bisisosorbide triglycerol

From the FTIR spectrum, the –OH group appearing at 3390 cm^{-1} as strong absorption, the –C-H group appearing as strong single-bond stretch at 2900 cm^{-1} and the C-O stretch appearing at 1100 cm^{-1}, consistent with the material being the desired compound. The NMR spectra further confirmed the structures.

The prepared isosorbide derived polyols were dried over phosphorus pentoxide to constant weight, then equilibrated at three relative humidities to measure their moisture uptakes. The different relative humidities (31%, 51%, 73% RH) were maintained by using calcium chloride, calcium nitrate, sodium nitrate respectively as saturated solutions at room temperature. A sample of polyethylene glycol (PEG 4000 MW) was included as a control. The results are recorded in Table I. As shown in the table, both isosorbide diglycerol and bisisosorbide triglycerol have a higher water absorption property than commercially available PEG at the three relative humidity conditions. Isosorbide diglycerol combined with more equivalent hydroxyl groups has a slightly higher water uptake ratio than bisisosorbide triglycerol. The abundant hydroxyl groups combined with hydroscopic isosorbide enhanced the water uptake properties of these polyols. As candidates for high water uptake compounds, this new class of value-added bio-polyols has the potential as moisturizers and humectants in cosmetic and personal care products. Since isosorbide can be chemically attached to glycerin in different ratios as a hydrophilic head, the free hydroxyl group on the glycerin could be esterified with different fatty acids, like lauric acid and palmitic acid, to form the non-ionic surfactants as alternatives to PEG based surfactants, where PEG is made by polymerizing ethylene oxide, a highly volatile and potential explosive intermediate. Meanwhile, the bisisosorbide triglycerol, whose glycerins are generated either before or after the acidic hydrolysis, can be multi-functionalized through chemical modification of its hydroxyl groups at different stages. For instance, the center hydroxyl group on the bisisosorbide diglycidyl ether could be alkylated first with fatty acid or alkyl halide to to generate the surfactant functionalized epoxide. Followed with acidic hydrolysis, the preliminary epoxide could be opened up to form the corresponding polyols, where the freshly prepared non-alkylated hydroxyl groups could be further attached to different functional moieties, such as UV-absorbing chromophores, or radical chelating antioxidants, to name a few. By using this technology, a variety of derivatives of isosorbide based polyols can be prepared as the new "green" small molecules and polymers with multi-functionalities for specialty chemical industry. Furthermore, when isosorbide derived polyols are attached to methyl methacrylates, isosorbide-glycidyl methacrylate (isoGMA) can be produced as prepolymer to replace Bisphenol A-glycidyl methacrylate (bisGMA) as a new generation dental composites for biomedical applications.

Isosorbide Based Thermoplastics

Isosorbide is generally known as a monomer for incorporation into polyesters such as polyethylene terephthalate (PET) at low levels (*38–41*). Due to the rigid molecular structure of isosorbide, polyester copolymers made by using isosorbide have higher glass transition temperatures (Tg) than PET, with the range from 80℃ to 200℃. The increasing heat resistance of copolymer depends on the incorporation ratio of isosorbide. The more the isosorbide monomer is incorporated, the higher the Tg of polyesters with more amorphous phase could be. Because there is growing interest in inexpensive bottle resins, which have a Tg of 86℃, or higher, and can be hot filled and pressurized without distortion, the low cost isosorbide reinforced polyester with attractive thermal stability becomes to be a very promising candidate for the application of "hot filling" bottles, like tomato ketchup or other condiments that must be pasteurized first. However, incorporation of isosorbide into PET on a commercial scale has encountered several problems. The secondary hydroxyl groups of isosorbide make it less reactive than the primary hydroxyls of ethylene glycol. This fact, coupled with the volatility of isosorbide, make it difficult to get high incorporation into (polyethylene isosorbide terephthalate) PEIT copolymers and leads to complications with the recycle of the ethylene glycol/isosorbide stream generated during polymerization (*42*). Additionally, the stereochemistry of isosorbide is such that one OH is ***endo*** and one ***exo***: this leads to a pronounced differential reactivity and an overall sluggishness in copolyester formations. Usually only 35-50% of the added isosorbide becomes part of the polymer structure – the rest is lost by evaporation in the high vacuum stage of the polymerization reaction.

In order to overcome these shortcomings, isosorbide derived AB monomers were proposed. The asymmetric synthesis allows isosorbide to be functionalized in the different stages. Benzyl ether acting as a protection group could be attached to one of the hydroxyl groups on isosorbide specifically. Through the Williamson ether reaction, a mixture of isosorbide derived 2- and 5-monobenzyl ethers generated by adding the reactive alkoxide to a benzyl chloride or benzyl bromide. The 2-monobenzyl ether **4** is easily separated by recrystallization from diethyl ether and the 5-monobenzyl ether **5** can be isolated by distillation of the mother liquors. With the benzyl ether protected isosorbide, different functional groups such as terephthalate can be attached to the other end of hydroxyl group. After hydrogenation, the desired isosorbide AB monomer has been prepared as shown in Scheme 6. The chemically modified isosorbide derivatives allow for essentially complete incorporation into high molecular weight polymers like PET and PLLA to raise their Tg. The homopolymers from these AB monomers also represent a new class of polymers in their own right.

Preparation of isosorbide 5-(4-carbomethoxybenzoate) (isosorbide 5-TA 7) is shown in Scheme 7. Isosorbide 2-benzyl ether 5-(4-carbomethoxybenzoate) was prepared first as pre-monomer before the hydrogenation. Freshly prepared 4-carbomethoxy-benzoyl chloride **6** was reacted with isosorbide 2-benzyl ether **4** to form the corresponding ester. Because of the high reactivity of the acyl halide, this reaction is often too vigorous to control and evolves excess heat, which results in a mixture of impure products. Therefore, this reaction should be kept cold, around 0℃. The benzyl ether protected pre-monomer was then hydrogenated at low pressure (1-5 atm) using Pd/C as a catalyst to yield isosorbide 5-TA 7 as the desired product with purity of 99.95%, mp 155℃.

Similar to isosorbide 5-TA 7, isosorbide 2-TA **9** was synthesized through the same procedure as shown in Scheme 8, where the starting material was switched from isosorbide 2-benzyl ether **4** to isosorbide 5-benzyl ether **5**. The product was 99.66% pure with mp. 143℃

Table I. Weight Percent Water Absorption at RH

Compound	*31% RH*	*51% RH*	*73% RH*
Isosorbide Diglycerol	0.7	7	23
Bisisosorbide Diglycerol	0.6	5.8	21
PEG(4000MW)	0.2	0.6	1.6

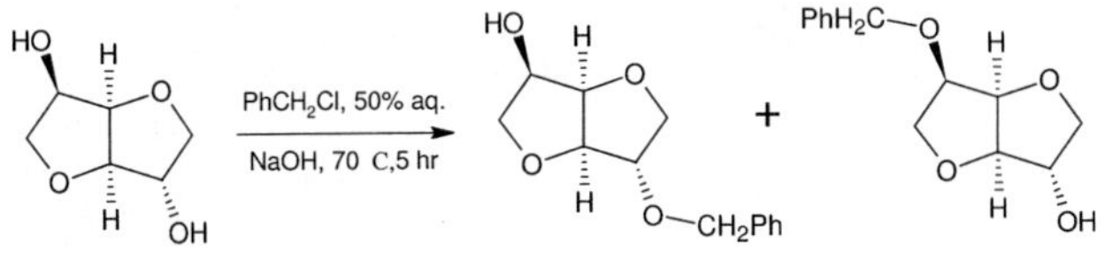

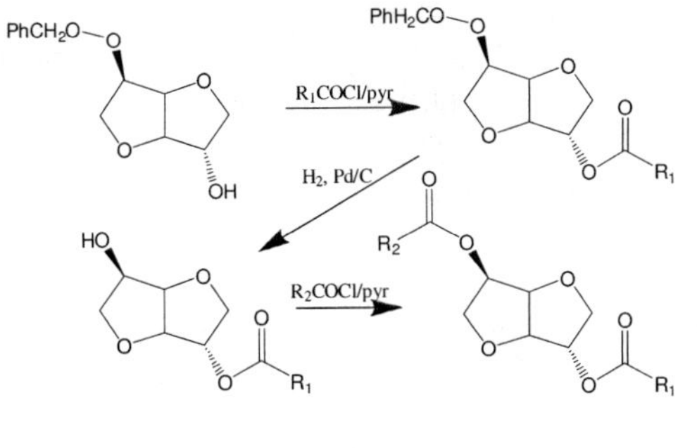

Scheme 6. Synthetic route for asymmetric synthesis

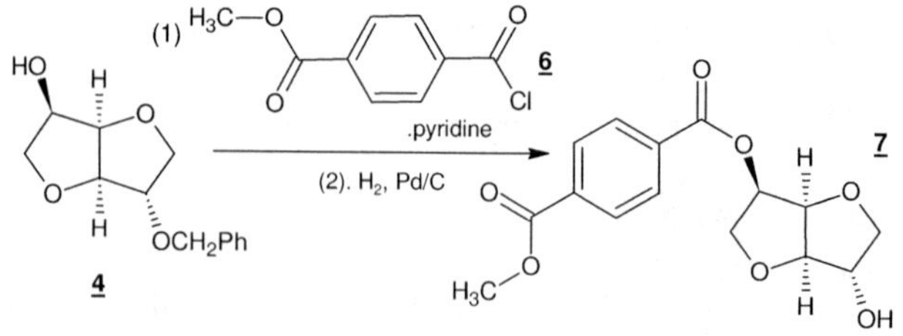

Scheme 7. *Synthetic route of isosorbide 5-TA*

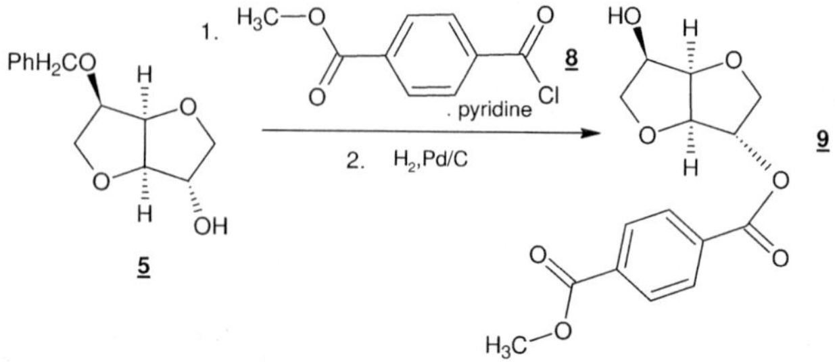

Scheme 8. *Synthetic route of isosorbide 2-TA*

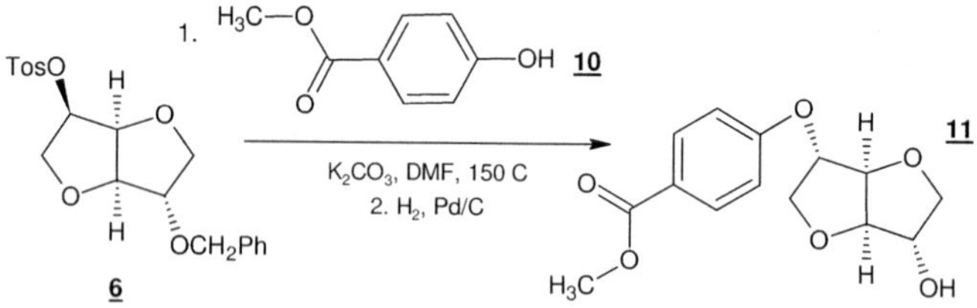

Scheme 9. *Synthetic route of isoidide 2-(4-carbomethoxyphenyl) ether*

For the isoidide derived AB monomer, the SN2 chemistry was used as shown in Scheme 9, where the isosorbide 2-benzyl ether 5-tosylate **6** was reacted with methyl 4-hydroxybenzoate **10** to invert the ***endo***-tosylate at the C-5 position to the ***exo***-ether. The original isosorbide was converted to its isomer, isoidide through the Walden Inversion reaction. After hydrogenation of the benzyl ether, the isoidide 2-(4-carbomethoxyphenyl) AB monomer was prepared with purity of 99.7%, mp. 117℃. Because the isoidide derived AB monomer contains the ether linkage, which should be chemically and thermally more stable than an ester linkage and cannot randomize during polymerization, a new class of polymers are expected as isoidide derived poly(ether-ester)s.

Purified monomers were melt-polymerized individually under vacuum up to 280℃ at increments of 20 ℃ / hour by using titanium isopropoxide as catalyst. NMR analysis of the resulting polymer in $CDCl_3$ indicated a degree of polymerization (DP) of ca. 10 giving Mn = 2500 polymers. Isosorbide based polyesters made from isosorbide 2-TA 9 and isosorbide 5-TA 7 showed a Tg around 145℃ without crystalline melting point. After annealing above their Tg for several hours, the DSC trace in both cases showed an increased Tg and crystalline melting endotherm, which is to be expected for semicrystalline polymers. For the poly(ether-ester) containing isoidide **11**, because of its rigid ether linkage, the degree of freedom is reduced for this type of polymer. Its Tg rose to 170 ℃ and the melting point appeared at ca. 300℃. After annealing, the crystallinity increased based on an increased heat of fusion. By using Thermogravimetric Analysis (TGA) from a TA Instrument thermal analyzer, the thermal stability was measured for these homopolymers. All of them are thermally stable below 320℃. Their NMR spectra showed they were stereo-consistent with their respective monomers. Because isosorbide 5-TA and 2-TA result in the same stereoregular structure when polymerized, their polymers gave very similar NMR spectra, as expected. The only difference between the two was their different rates of polymerization, since one has an ***exo*** hydroxyl group and the other has an ***endo*** hydroxyl group.

Based on Ballauf's previous work, when isosorbide was polymerized as a diol with terephthaloyl chloride, the resulting polymer was an amorphous material with randomly scrambled orientation as shown in scheme 10 (*43*, *44*). The regio-irregular isosorbide provides two possible incorporation orders compared with symmetric isomannide and isoidide, which make semicrystalline materials when polymerized with TA unit. Since each of isosorbide derived AB monomers only offers one incorporation order, the polyesters generated from them are regioregular with controlled stereochemistry as shown in scheme 11. The existence of crystallinity in their homopolymers demonstrates their potential applications in hot fill packaging, in which crystallites are required to stabilize bottles during the hot filling process without deforming. The AB monomers incorporate larger molar fractions of isosorbide into polyesters without forming amorphous material. Based on Flory-Fox equation, the Tg of PEIT can be raised from 70℃ to 150℃, and even 170℃ in the special case of the isoidide derived ether AB monomers. Since each single monomer has 1:1 ratio of carbonyl group to hydroxyl group, AB homopolymers can provide controlled stoichiometry of the polymerization process. As isosorbide AB monomer is molecularly bulky and pure, it should reduce sublimation and reduce the color production in melt polymerizations, which is important for making bottles, hot-fill containers, films, sheets, fibers, strands and optical articles.

With improved thermal stability, isosorbide AB monomer could be incorporated into commercial available polymers like PET and PLLA by mixing them with PET or PLLA monomers followed by polymerization. Alternatively, transesterification reaction with preformed PET or PLLA and AB monomers or their oligomers will likewise give a copolymer. Polycondensation would provide valuable polymeric materials, such as polyesters, polyamide esters, polyurethanes and polycarbonates which incorporate isosorbide AB monomers, could show better thermal stability than the isosorbide polyesters of the known art. The homopolymers themself as a new class of polyesters also present new opportunities for high use temperature polymers with Tg higher than 150℃ and melting point higher than 300℃. Because of the rigidity and chirality, AB monomers should make cholesteric liquid crystal polymers (LCP) with different properties compared with naphthalene based LCPs.

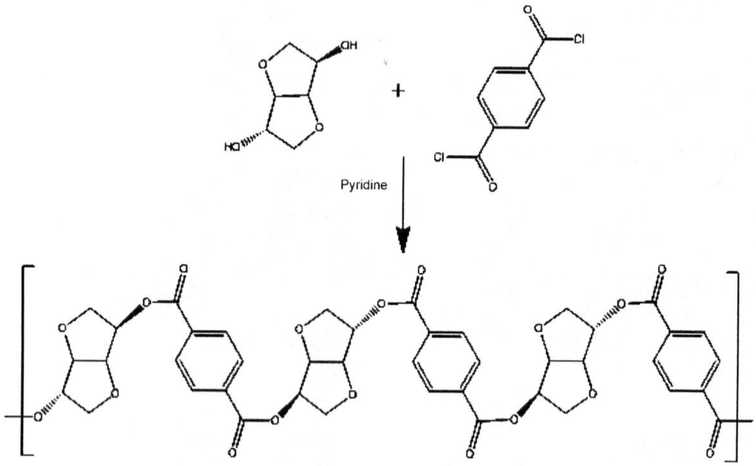

Pyridine

Scheme 10. Synthesis of PEIT with random orientation of Isosorbide

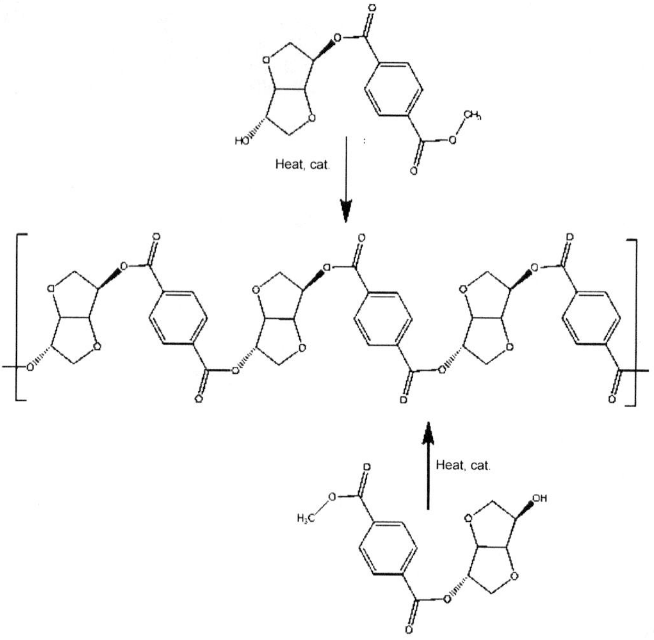

Heat, cat.

Heat, cat.

Scheme 11. Synthesis of stereoregular PEIT from AB monomers (7 and 9)

Isosorbide Based Designer Compounds

As described above, isosorbide is a glucose-derived natural material. It has found use in the form of derivatives in the pharmaceutical industry like isosorbide mononitrate and isosorbide dimethyl ether. Since isosorbide is water soluble and harmless, it can be used as an intermediate for additives and stabilizers in both cosmetics and plastics industry. Depending on the different reactivity of its two hydroxyl groups, isosorbide acts as a nucleus allowing different functionalities to be attached, such as UV-absorbing moieties, antioxidants and plasticizers. Cinnamic acids are widespread in nature and act as natural UV absorbers: examples are 4-methoxycinnamic acid and ferulic acid. A synthetic derivative of α-cyano-ferulic acid was also examined as a UVA long-wavelength absorbing moiety; methyl 3-(4-hydroxy-3, 5-di-t-butylphenyl) propionate and ferulic acid were used as antioxidant moieties. The two isomeric monobenzyl ethers of isosorbide both act as plasticizers for PVC and can be made selectively as protective groups for the hydroxyl group under appropriate conditions as described above. Starting from a benzylether protected isosorbide, different functional groups like 4-methoxy-cinnamic acid can react with the other hydroxyl group to produce a UV absorbing plasticizer. Hydrogenolysis of the benzyl ether enables one to attach a second functionality such as an antioxidant or a new UV absorber (e.g. UV-A), provided the first group is stable to hydrogenation. Full-spectrum UV absorbers and UV absorber-antioxidants can be synthesized.

Isosorbide Derived UV Absorber

Based on the data of The National Institutes of Health, skin cancer, the most common cancer in the USA, affects more than 1,000,000 Americans every year. It accounts for more than 10,000 deaths annually (*45*). Since 1983, various studies have revealed that chronic sun exposure plays an important role in the etiology of skin cancers, inflammation and skin aging (*46–48*).

The electromagnetic spectrum emitted by the sun ranges from the very short wavelength cosmic rays to the very long wavelength radio waves (*49*). The ultraviolet (UV) rays which comprise the shortest of the non-ionizing rays are responsible for most of the photo-cutaneous changes. High amounts of energy in photons of ultraviolet radiation can be absorbed by the photosensitizing molecules in the body causing degradation, chemical reaction and the generation of toxic radicals. The shorter the wavelength, the more damaging is caused by the UV radiation. The damaging UV radiation is usually divided into four categories: UVC rays (100~290nm), UVB rays (290~320nm), UVAII rays (320~340nm) and UVAI rays (340~400nm). UVC radiation is completely absorbed by the atmosphere before it reaches the ground. UVB and UVA rays can penetrate through the atmospheric ozone layer and trigger photobiological events in the skin.

UVB radiation with shorter wavelength is normally absorbed by the epidermis of skin. It is the primary carninogenic stimulus for skin cancers through clinical formation of pyrimidine dimmers (*50*). These dimers cause lesions on the DNA double strand, which arrest the replication of DNA by inhibiting their enzyme polymerases. This process is often called the direct DNA damage (*51*). Normally this damage is reduced to some extent due to the excellent photochemical properties of DNA. DNA can transform more than 99.9% of the photons into harmless heat through ultrafast internal conversion process (*52*). Similar to the DNA, melanin which is generated by melanocytes in the epidermis also acts as a defender to UVB radiations. It has extremely short excited state lifetimes in the range of a few femtoseconds (10^{-15}s) after absorbing the high energy UVB light (*53*). This short excited state lifetime forbids melanins from reacting with other biological molecules and only allows them to transform the photo energy into the harmless heat causing sunburn and pigmentation changes as a warning signal (*54*). Even though this mechanism provides a good self-defense system to the body, it only can sustain itself for a short time. Chronic UVB radiation still can produce squamous cell carcinoma in animals and human skins (*55, 56*).

Compared with UVB rays, which do not pass through window glass, UVA rays with lower energy **do** pass through window glass and even penetrate into the deeper skin layers to produce a series of phototoxic events in the skin. UVA rays can trigger photosensitive molecules in the dermis into their lowest excited state, where the excited molecules have a long enough half-life to undergo photochemical reactions (*57*). These excited molecules could either damage the living cells which come into contact with them or react with biological species in the body to generate free radicals or reactive oxygen species. Since these reactive species including singlet oxygen, hydrogen peroxide, and superoxide anion are diffusible in the extracellular matrix, they can oxidize lipids and proteins to initiate matrix metalloproteinases, which destroy the connective tissue and result in photoaging of skin. They can also travel into the cells and attack DNA molecules to cause indirect DNA damage by causing strand breaks and oxidizing purine (A or G) nucleic acids (*58, 59*). Since this type of DNA damage can be amplified through persistent radical reactions, it enables malignant melanoma to occurr in places that are not directly illuminated by the sun (*60*). Since UVA rays can reach the earth with significant amount, increasing attention should be paid to protection from UVA radiation, especially to their carcinogenesis, photoaging, and immunosuppression effects (*61*).

Increasing awareness of these damaging effects of sunlight have let to a significant demand for improved photoprotection by topical applied sunscreen agents. In the United States, the Food and Drug Administration (FDA) regulates sunscreen products as over-the-counter (OTC) drugs, which leads the less frequent introduction of new sunscreen actives compared with in Europe (*62, 63*). Today there are twenty-eight and sixteen sunscreen ingredients approved in Europe and the United States respectively (*64*). Based on their different working mechanisms, sunscreens can be categorized into two types: reflection sunscreen and absorption sunscreen. For reflection sunscreens, since scattering of UV light is influenced by the difference in refractive index between the pigment and the surrounding media, inorganic material with high refractive index becomes an optimal choice to generate reflection of UV radiations. The most common inorganic sunscreens include TiO_2 and ZnO. ZnO could reflect almost all the UVB and UVA rays (250-380nm). TiO_2 reflects UVB and UVAII regions (250-240nm). The reflection capacity of inorganic sunscreens is decided by their two parameters: 1) refractive index; 2) particle size. Based on the Mie theory, the maximum light scattering occurs when the size of particles equals 1/2 the wavelength of electromagnetic radiation. Because most of traditional TiO_2 and ZnO materials are bigger than 200nm, they will reflect visible light with wavelength of greater than 400nm. This type of sunscreens appears white on the skin. New technologies can make materials with particle sizes smaller than 100nm so that they will not scatter visible light and become transparent. But these inorganic sunscreens agglomerate and begin appearing white at higher concentrations. Good formulation technology is required for this type of sunscreen. Since these inorganic sunscreens can generate highly oxidizing radicals and other reactive oxygen species (ROS) such as H_2O_2 and singlet oxygen to produce cytotoxic and/or genotoxic effects on skin cells, there is still a controversy on the safety issues of inorganic sunscreens (*65–67*).

Compared with inorganic sunscreen, organic sunscreens make up 80% of the commercial market. The organic sunscreens are less expensive to produce and are easier to formulate than inorganic blockers. Organic sunscreens normally include a chromophore to absorb high-intensity UV rays. Generally, this chromophore comprises an extended conjugation system, such as an aromatic ring conjugated with a carbonyl group (*68*). Because this conjugated structure can reduce the energy difference of compound between its excited state and ground state, when its energy difference equals to the energy of UV radiation, the compound is able to absorb corresponding UV rays. The more conjugation is involved in the compound, the longer the UV rays with lower energy that can be absorbed by the chromophore. After absorbing the UV radiation, ideally, the excited chromophores should transfer the energy of UV radiation to the molecular thermal motions such as vibration and rotation of molecules, and finally reemit it as heat or harmless long wave IR. However, not all of chromophores can undergo this process. Some excited chromophores undergo chemical reaction such as photodimerization or photoisomerization thus lose their original sun scanning functions during sun exposures.

Designing a sunscreen is not easy. Several requirements have to been satisfied for both inorganic and organic sunscreens, such as photo-stable, human friendly and broadband UV absorptions. Today, not many approved sunscreens possess these properties of improved photoprotection. The major opportunity here for isosorbide is its non-toxicity and sustainability. Natural derived cinnamic acid as chromophore can be stabilized by isosorbide to provide consistent UV absorption property. By adding different functionalities onto isosorbide, several multifunctional sunscreens can be generated as a whole new class of UV absorbers.

The preparation of isosorbide based sunscreens was done through esterification reactions, where an ester linkage was formed between isosorbide and various functionalized cinnamic acids. When two 4-methoxycinnamic acids were attached to the two sides of isosorbide under the slight acidic condition with a p-toluenesulfonic acid catalyst, the isosorbide 2,5-bis(4-methoxycinnamate) was prepared as the first isosorbide derived UVB sunscreen. The synthetic route is shown in the Scheme 12 and the starting material 4-methoxycinnamic acid was prepared by a Knoevenagel reaction of 4-methoxybenzaldehyde with malonic acid in pyridine with a 4-methylpiperidine catalyst. Following the same method, isosorbide bis(3,4,5-trimethoxycinnamate) and isosorbide bisferulate were also synthesized.

For isosorbide based UVA sunscreens, appropriate chemical modification of cinnamic acids and esterifying them onto isosorbide makes the UV absorption of the compounds shift from short wavelength UVB region to the long wavelength UVA region. Different absorbances in the UVA region are expected for the different modified UVA sunscreens. As shown in the scheme 13, the isosorbide biscyanoferulate as the first isosorbide derived UVA sunscreen was prepared through refluxing isosorbide biscyanoacetate with vanillin using a 4-methylpiperidine-acetate salt catalyst in benzene. The isosorbide biscyanoacetate was prepared by heating ethyl cyanoacetate with isosorbide using the dibutyltin dilaurate as catalyst under N_2 flow. Isosorbide bis(3, 4-dimethoxy-cyanocinnamate) and isosorbide bis(3, 5-dimethoxy-4-hydroxy-cyanocinnamate) were also prepared following the same scheme.

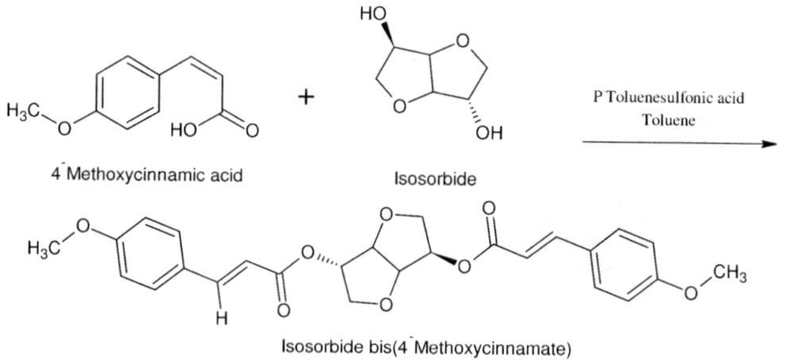

Scheme 12. Synthetic route of isosorbide bis(4-methoxycinnamate)

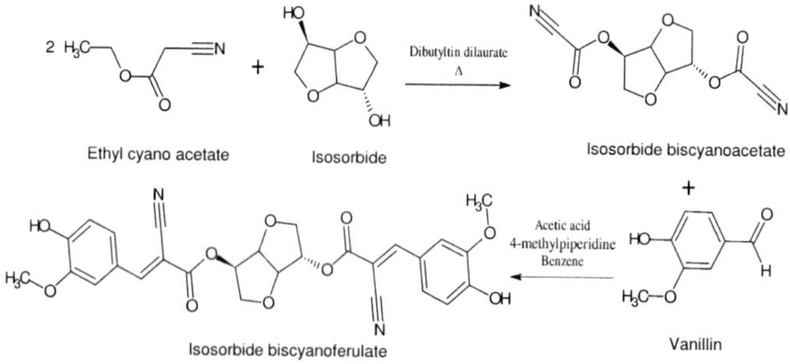

Scheme 13. Synthetic route of isosorbide biscyanoferulate

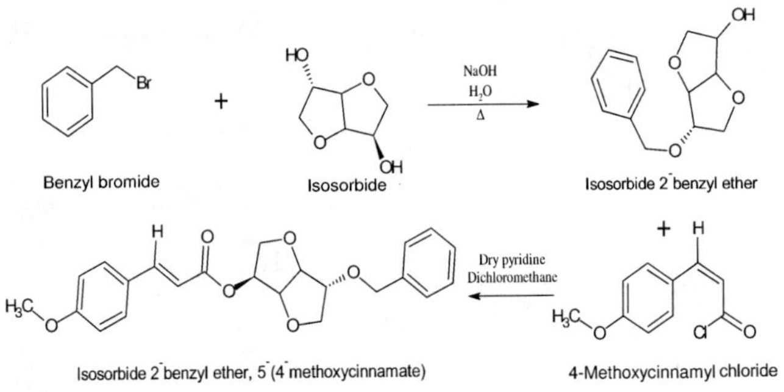

Scheme 14. Synthetic route of isosorbide 2-benzyl ether, 5-(4-methoxycinnamate)

For multifunctional sunscreens, based on the different reactivity of hydroxyl groups on isosorbide, different functional groups could be added on each side of isosorbide at different stages. UVB and UVA absorbers could be added to form broadband UV absorbers. UV absorption and plasticizer could be added to form UV absorbing plasticizers as multifunctional additives in the polymer resins, where the UV light can break down the chemical bonds in the polymer and ultimately causing the cracking, chalking, color changes and loss of physical properties such as tensile strength, impact strength, elongation and others. Taking the isosorbide derived UV absorbing plasticizer as an example, it was prepared as shown in Scheme 14. The isosorbide 2-benzylether was prepared by refluxing isosorbide with benzyl bromide and 50% sodium hydroxide solution for 8 hours. The product which was precipitated with diethyl ether and recrystallized from MTBE was acylated in dry pyridine and dichloromethane with 4-methoxycinnamoyl chloride at 5℃.

Protons of every compound were assigned in the ^1H NMR ($CDCl_3$) spectrum to confirm their correct structures. Taking isosorbide bis(4-methoxycinnamate) as an example (Figure 2), the 1:1 integrated ratio of H(2) and H(5) at 5.332 and 5.303 ppm demonstrates that the esterification reaction took place on both hydroxyl groups of isosorbide. The H(8) and H(8′) appear at 7.7ppm as two sets of doublets overlapping with each other due to the coupling with H(7) and being next to the asymmetric system of isosorbide. For H(7) and H′(7), since they are more close to the asymmetric isosorbide, the difference between their chemical shift appears larger. The 16 Hz coupling constant for both H(7) and H(8) demonstrates the trans configuration of their double bond. Coupled with H(9) and H′(9), H(10) and H′(10) are shown at 6.9 ppm as a doublet. Since H(9) and H′(9) are influenced more by asymmetric isosorbide, they appear at 7.3ppm as complexity of peaks. H(4) on the isosorbide coupled with H(3) and H(5) appears at 4.9ppm as a triplet. H(3) coupled with H(4) shows up at 4.6ppm as a doublet. Six protons of H(11) and H′(11) in the 3.8 ppm region are shown as two singlet with the influence of asymmetric isosorbide.

For the UVA sunscreens, an ATI MATTSON FTIR spectrophotometer was used to detect their modified functional group. Taking isosorbide biscyanoferulate as an example (Figure 3), the –OH group appears at 3370 cm^{-1} as strong absorption; the –CN group appears as sharp triple-bond stretch at 2220 cm^{-1} and the C=O group of the ester appears at 1740 cm^{-1}, which illustrates that the material has maintained all the desired functional groups.

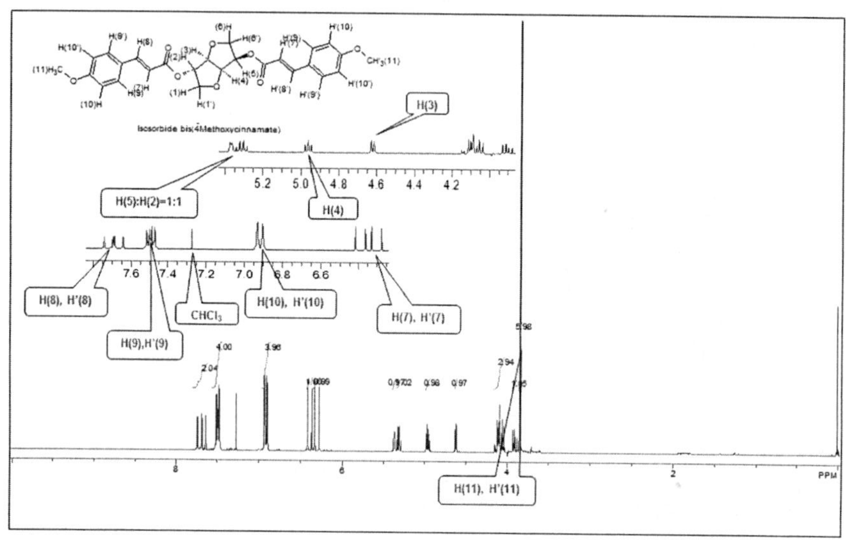

Figure 2. NMR spectrum of isosorbide bis(4-methoxycinnamate)

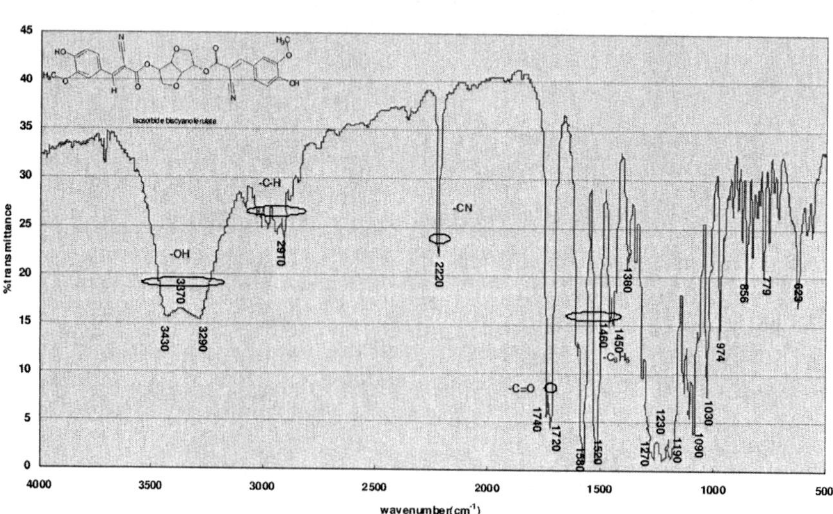

Figure 3. FTIR spectrum of isosorbide biscyanoferulate

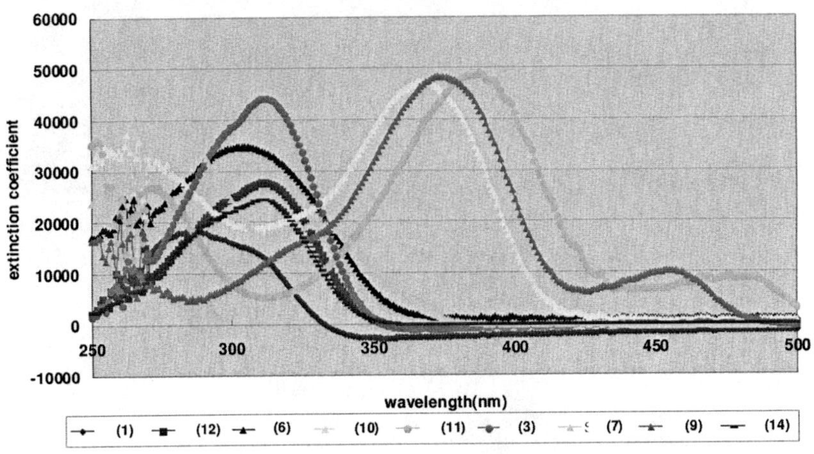

Figure 4. UV spectrum of all synthetic sunscreen: 4-methoxycinnamic acid (1); Isosorbide 2-(4-methoxycinnamate) (12); Isosorbide Bis-(4-methoxycinnamate) (3); Isosorbide Bis(3,4,5-trimethoxycinnamate) (6); Isosorbide Bisferulate (7); Isosorbide Bis(3,4-dimethoxycyanocinnamate) (10); Isosorbide Biscyano (3,5-dimethoxy, 4-hydroxycinnamate) (11); isosorbide biscyanoferulate (9); Isosorbide 2-(benzyl ether)- 5-(4-methoxycinnamte)(14).

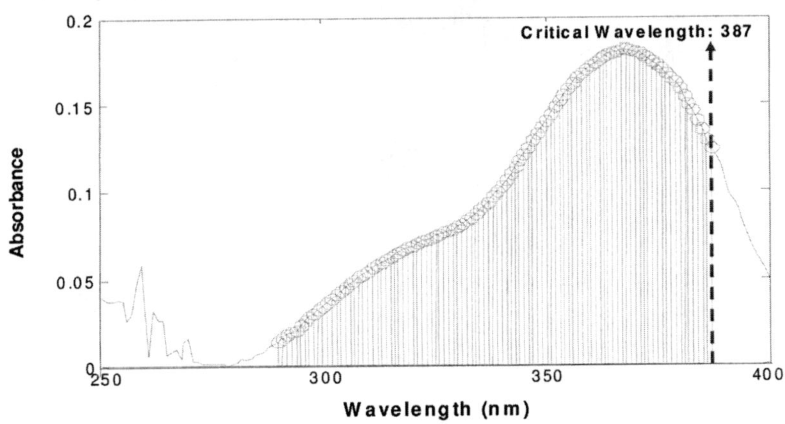

Figure 5. Critical wavelength of isosorbide (3,4-dimethoxycyanocinnamate)

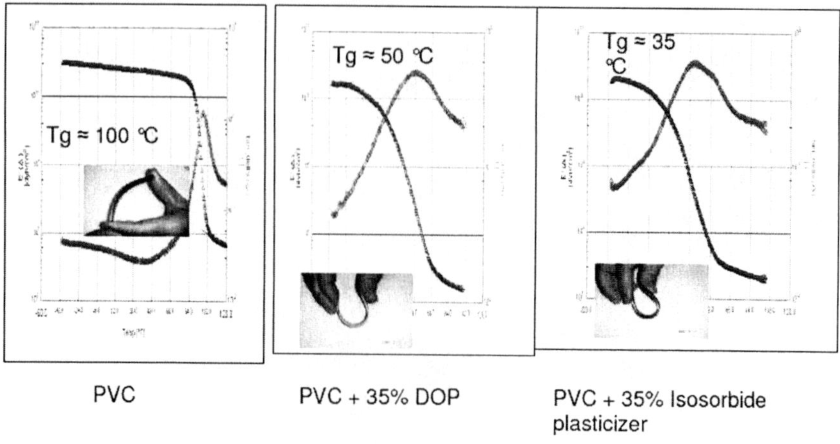

Figure 6. DMTA of Isosorbide Based Plasticizer

The UV spectra of several isosorbide derived sunscreens are shown in the Figure 4. Isosorbide as an intermediate in these sunscreens connects two chromophores together without disturbing their UV absorption properties. The combination of the two chromophores provides the high extinction coefficient. Due to the electron withdrawing effect, the cyano group on a modified cinnamate causes a bathochromic shift for the UVA chromaphore. The combination of isosorbide derived sunscreens covers the whole UVA/UVB range with high molar extinction coefficient.

In order to characterize the long-wavelength UVA property, Diffey's method was used to test the synthetic UVA sunscreens, where critical wavelength (λ_C) was measured at which the integral of the spectral absorbance reached 90% of the integral from 290 to 400nm (*69*). Sunscreens with critical wavelength greater than 370nm could be claimed as longer wavelength UVA covering sunscreens. Taking Isosorbide bis (3, 4-dimethoxy-cyanocinnamate) as an example, its critical wavelength was calculated based on the equation:

$$\int_{290}^{\lambda_c} A(\lambda)\,d\lambda = 0.9 \int_{290}^{400} A(\lambda)\,d\lambda$$

As shown in figure 5, the λ_C of 387nm demonstrated its long-wavelength UVA property. Since all the isosorbide-derived UVA sunscreens had critical wavelength greater than 370nm, they could all be claimed as sunscreens covering the longer wavelength part of the UVA range (340-400 nm).

The photostability test of the UVA sunscreens was tested at room temperature in an enclosed conatiner using a single focusing UV LED lamp (A160-010 from UVPS) irradiating at 375nm with energy of 200mW/cm^2. The samples in ethanol were analyzed by UV/Visible spectrometry before and after an 8 hour irradiation. The Area Under the Curve Index (AUCI) was calculated as the ratio of AUC_{after}/ AUC_{before} based on the change of UV spectra. The AUCI for three UVA sunscreens was > 80%, which illustrates they are photo-stable in ethanol for 8 hours, the normal average exposure time of humans in sunlight.

From all the results discussed above, the conclusion could be made: as an effective alternative to petroleum based compounds, the cinnamic esters of isosorbide provide an environmentally sustainable strategy for preparing a whole new class of photo-stable and broad effective UV absorbers with multifunctionalities for both cosmetic and plastic industry.

Isosorbide Derived Plasticizer

Plasticizers are polymer additives that act as internal lubricants, increasing the fluidity of the material to which they are added. 80% of Plasticizers are used in the rigid and brittle PVC polymers (*70*). Phthalates as the most common plasticizers account for 92% of all plasticizers in the market (*71*). They are produced at about one million metric tons per year in Europe. It is now believed that phthalate based plasticizers can act as endocrine disrupters and should not be used in consumer products. To avoid the large exposure of children and infants to phthalates, new green alternatives are needed in both the food packaging and toy industries. The alkylated aromatic esters and ethers of isosorbide as naturally derived materials have been explored as effective plasticizers for PVC and other polymers of commercial interest. As shown in Figure 6, the dynamic mechanical response of molded bars of PVC with standard 35% PVC/DOP is compared with an isosorbide based plasticizer composition. The isosorbide based plasticizer provides comparable performance to the dioctylphthalates (DOP) in their ability to plasticize PVC. The isosorbide blend PVC shows a lower Tg than the DOP blend PVC, which demonstrates that the isosorbide compound is as effective as DOP and even better than DOP at similar additive levels.

Conclusion

The polymer industry represents a large volume opportunity for isosorbide based low molar mass designer compounds, monomers, and high molar mass polymers. Sugar derived chemicals especially isosorbide with a built-in chirality and stereochemistry provides a ubiquitous platform for environmentally sustainable chemicals in both the plastics and cosmetics industry. Following the concept of cost effectiveness, value added products have been designed based on the asymmetric reactivity of isosorbide. Varies applications for isosorbide, including environmental friendly epoxides, high water uptake humectants, heat resistant thermoplastics and multifunctional additives, stand as proof of the bright future of sugar based chemistries. Their excellent performance suggest the possibility of their commercial success. As petroleum based products becomes more expensive and the pressure mounts to create sustainable products, the creation of alternative chemistries from renewable resources such a sugar becomes more attractive.

References

1. Fernando, S.; Adhikari, S.; Chandrapal, C.; Murali, N. Biorefineries: Current status, challenges, and future direction. *Energy Fuels* **2006**, *20*, 1727–1737.
2. Levy, P. F.; Sanderson, J. E.; Kispert, R. G.; Wise, D. L. Biorefining of biomass to liquid fuels and organic chemicals. *Enzyme Microb. Technol.* **1981**, *3*, 207–215.
3. Elliott, E. *Polymers and People: An Informal History*; Beckman Center for the History of Chemistry: Philadelphia, PA, 2005.
4. Stevens, M. P. *Polymer Chemistry: An Introduction*, 2nd ed.; Oxford University Press; New York, 1999.
5. Morris, P. J. T. *Polymer Pioneers: A Popular History of the Science and Technology of Large Molecules*; Center for History of Chemistry: Philadelphia, PA, 2005.
6. Van Haveren, J.; Oostveen, E.; Miccichè, F.; Noordover, B.; Koning, C.; van Benthem, R.; Frissen, A.; Weijnen, J. Resins and additives for powder coatings and alkyd paints, based on renewable resources. *J. Coat. Technol. Res.* **2007**, *4*, 177–186.
7. Willke, T.; Vorlop, K. D. Industrial bioconversion of renewable resources as an alternative to conventional chemistry. *Appl. Microbiol. Biotechnol.* **2004**, *66*, 131–142.
8. Azapagic, A.; Emsley, A.; Hamerton, L. *Polymers: The Environment and Sustainable Development*; John Wiley & Sons, Ltd.: West Sussex, England, 2003.
9. DeSimone, L. D.; Popoff, F. *Eco-Efficiency: The Business Link to Sustainable Development*, The MIT Press: Boston, MA, 2000.
10. Fukushima, K.; Kimura, Y. Stereocomplexed polylactides (Neo-PLA) as high-performance bio-based polymers: their formation, properties, and application. *Polym. Int.* **2006**, *55*, 626–642.
11. Mohanty, A. K.; Misra, M.; Drzal, L. T. *Natural Fibers, Biopolymers, and Biocomposites*; CRC Press: Boca Raton, FL, 2005.
12. Mohanty, A. K.; Misra, M.; Hinrichsen, G. Biofibres, biodegradable polymers and biocomposites: An overview. *Macromol. Mater. Eng.* **2000**, 276–277, 1–24.
13. *From One Earth to One World: An Overview by the World Commission on Environment and Development*; Oxford University Press: New York, 1987.
14. Lange, J. P. Sustainable development: Efficiency and recycling in chemicals manufacturing. *Green Chem.* **2002**, 546–550.
15. Georgopoulos, S. T.; Tarantili, P. A.; Avgerinos, E.; Andreopoulos, A. G.; Koukios, E. G. Thermoplastic polymers reinforced with fibrous agricultural residues. *Polym. Degrad. Stab.* **2005**, *90*, 303–312.
16. Gontard, N.; Guilbert, S. *Food Packaging and Preservation*; Springer: New York, 1994.
17. Narayan, R. In *Polymers from Agricultural Coproducts*; ACS Symposium Series 575; American Chemical Society: Washington, DC, 1994.
18. Scott, G. 'Green' polymers. *Polym. Degrad. Stab.* **2000**, *68*, 1–7.

19. Wool, R. P.; Sun, X. S. *Bio-Based Polymers and Composites*; Elsevier Academic Press: Boston, MA, 2005.
20. Legendre, B. L.; Burner, D. M. Biomass production of sugarcane cultivars and early-generation hybrids. *Biomass Bioenergy* **1995**, *8*, 55–61.
21. Godshall, M. A. *Future Directions for the Sugar Industry*; Sugar Processing Research Institute, Inc.: New Orleans, LA, 2001.
22. Chatti, S.; Bortolussi, M.; Loupy, A. Synthesis of new diols derived from dianhydrohexitols ethers under microwave-assisted phase transfer catalysis. *Tetrahedron* **2000**, *56*, 5877–5883.
23. Chatti, S.; Bortolussi, M.; Loupy, A. Synthesis of diethers derived from dianhydrohexitols by phase transfer catalysis under microwave. *Tetrahedron Lett.* **2000**, *41*, 3367–3370.
24. Malhotra, S. V.; Kumar, V.; East, A.; Jaffe, M. *Frontiers of Engineering: Reports on Leading-Edge Engineering*; National Academies Press: Washington, DC, 2008.
25. Stross, P.; Hemmer, R. 1,4:3,6-Dianhydrohexitols. *Adv. Carbohydr. Chem. Biochem.* **1991**, *49*, 93–175.
26. Wolfrom, M. L. *Advances in Carbohydrate Chemistry*, Academic Press: New York, 1955.
27. Cope, A. C.; Shen, T. Y. The Stereochemistry of 1,4: 3,6-dianhydrohexitol derivatives. *J. Am. Chem. Soc.* **1956**, *78*, 3177–3182.
28. Xianhong, F.; Jaffe, M.; Hammond, W. B.; East, A. J. Synthesis of Corn-Derived Carbohydrate Derivatives as Effective Multifunctional Sunscreens, Bioengineering Conference, 2009 IEEE 35th Annual Northeast, 2009, pp 1–2.
29. Fenouillot, F.; Rousseau, A.; Colomines, G.; Saint-Loup, R.; Pascault, J. P. Polymers from renewable 1,4:3,6-dianhydrohexitols (isosorbide, isomannide and isoidide): A review. *Prog. Polym. Sci.* **2010**, *35*, 578–622.
30. Brandner, J. D. U.S. Patent 3,023,223, Atlas Chemical Industries, 1962.
31. Mills, J. A. The Stereochemistry of Cyclic Derivatives of Carbohydrates. In *Advances in Carbohydrate Chemistry*; Academic Press: New York, 1955; Vol. 10, pp 1–53.
32. Feldmann, J.; Koebernick, H.; Richter, K.; Woelkin, H.. U.S. Patent 4,564,692, CPC International, Inc., 1986.
33. East, A., Jaffe, M., Zhang, Y., and Catalani, L.. Thermoset Epoxy Polymers from Renewable Resources. U.S. Patent Application 20080009599, New Jersey Institute of Technology, 2008.
34. Steinmetz, R.; Mitchner, N. A.; Grant, A.; Allen, D. L.; Bigsby, R. M.; Ben-Jonathan, N. The xenoestrogen bisphenol A induces growth, differentiation, and c-fos gene expression in the female reproductive tract. *Endocrinology* **1998**, *139*, 2741–2747.
35. Krishnan, A.; Stathis, P.; Permuth, S.; Tokes, L.; Feldman, D. Bisphenol-A: An estrogenic substance is released from polycarbonate flasks during autoclaving. *Endocrinology* **1993**, *132*, 2279–2286.
36. Staples, C. A.; Dome, P. B.; Klecka, G. M.; Oblock, S. T.; Harris, L. R. A review of the environmental fate, effects, and exposures of bisphenol A. *Chemosphere* **1998**, *36*, 2149–2173.

37. Xianhong, F.; DeMartino, R.; East, A. J.; Hammond, W. B.; Jaffe, M. Synthesis and Characterization of Isosorbide Derived Polyols as Highly Effective Humectants, Bioengineering Conference, Proceedings of the 2010 IEEE 36th Annual Northeast, 2010, pp 1–2.
38. Bengs, H.; Schoenfeld, A.; Boehm, G.; Weis, S.; Clauss, J. U.S. Patent 6,342,300 B1, Celanese Ventures GmbH, 2002.
39. Kricheldorf, H. R. "Sugar Diols" as building blocks of polycondensates. *Polym. Rev.* **1997**, *37*, 599–631.
40. Kricheldorf, H. R.; Behnken, G.; Sell, M. Influence of isosorbide on glass-transition temperature and crystallinity of poly(butylene terephthalate). *J. Macromol. Sci., Part A: Pure Appl. Chem.* **2007**, *44*, 679–684.
41. Thiem, J.; Lüders, H. Synthesis of polyterephthalates derived from dianhydrohexitols. *Polym. Bull.* **1984**, *11*, 365–369.
42. Hayes, R.; Brandenburg, C. U.S. Patent 6,608,167 B1, E. I. du Pont de Nemours and Company, 2003.
43. Storbeck, R.; Ballauff, M. Synthesis and thermal analysis of copolyesters deriving from 1,4:3,6-dianhydrosorbitol, ethylene glycol, and terephthalic acid. *J. Appl. Polym. Sci.* **1996**, *59*, 1199.
44. Storbeck, R; Rehahn, M; Ballauff, M. Synthesis and properties of polyesters based on 2,5-furandicarboxylic acid and 1,4:3,6-dianhydrohexitols. *Makromol. Chem.* **1993**, *194*, 53–62.
45. *Cancer Facts and Figures 2008*; American Cancer Society, 2008.
46. Kligman, L. H.; Akin, F. J.; Kligman, A. M. The contributions of UVA and UVB to connective tissue damage in hairless mice. *J. Invest. Dermatol.* **1985**, *84*, 272–276.
47. Gilchrest, B. A. Skin aging and photoaging: An overview. *J. Am. Acad. Dermatol.* **1989**, *21*, 610–613.
48. Epstein, J. Photomedicine. In *The Science of Photobiology*; CRC Press: Boca Ratton, FL, 1989; pp 155–192.
49. Biswas, S.; Fichte, C. E. Composition of solar cosmic rays. *Space Sci. Rev.* **1965**, *4*, 709–736.
50. Lowe, N. J.; Shaath, N. A.; Pathak, M. A. *Sunscreens: Development, Evaluation, and Regulatory Aspects*; Marcel Dekker, Inc.: New York, 1997.
51. Peak, M. J.; Peak, J. G. *Effects of Solar Ultraviolet Photons on Mammalian Cell DNA*; Argonne National Laboratory: Argonne, IL, 1991.
52. Pecourt, J. L.; Peon, J.; Kohler, B. DNA excited-state dynamics: Ultrafast internal conversion and vibrational cooling in a series of nucleosides. *J. Am. Chem. Soc.* **2001**, *123*, 10370–10378.
53. Crespo-Hernández, C. E.; de La Harpe, K.; Kohler, B. Ground-state recovery following uv excitation is much slower in G·C–DNA duplexes and hairpins than in mononucleotides. *J. Am. Chem. Soc.* **2008**, *130*, 10844–10845.
54. Parrish, J. A.; Jaenicke, K. F.; Anderson, R. R. Erythema and melanogenesis action spectra of normal human skin. *Photochem. Photobiol.* **2008**, *36*, 187–191.
55. Black, H. S.; DeGruijl, F. R.; Forbes, P. D.; Cleaver, J. E.; Ananthaswamy, H. N.; deFabo, E. C.; Ullrich, S. E.; Tyrrell, R. M. Photocarcinogenesis: An overview. *J. Photochem. Photobiol., B* **1997**, *40*, 29–47.

56. Nomura, T.; Nakajima, H.; Hongyo, T.; Taniguchi, E.; Fukuda, K.; Li, L. Y.; Kurooka, M.; Sutoh, K.; Hande, P. M.; Kawaguchi, T.; Ueda, M.; Takatera, H. Induction of cancer, actinic keratosis, and specific p53 mutations by UVB light in human skin maintained in severe combined immunodeficient mice. *Cancer Res.* **1997**, *57*, 2081–2084.
57. Cantrell, A.; McGarvey, D. J. Sun protection in man. *Compr. Ser. Photosci.* **2001**, *495*, 497–519.
58. Tyrrell, R. M. Oxidant, antioxidant status and photocarcinogenesis: The role of gene activation. *Photochem. Photobiol.* **2008**, *63*, 380–383.
59. Wenczl, E.; Pool, S.; Timmerman, A. J.; Schans, G. P.; Roza, L.; Schothorst, A. A. Physiological doses of ultraviolet irradiation induce DNA strand breaks in cultured human melanocytes, as detected by means of an immunochemical assay. *Photochem. Photobiol.* **1997**, *66*, 826–830.
60. Moan, J.; Dahlback, A.; Setlow, R. B. Epidemiological support for an hypothesis for melanoma induction indicating a role for UVA radiation. *Photochem. Photobiol.* **1999**, *70*, 243–247.
61. Ong, C. S.; Keogh, A. M.; Kossard, S.; Macdonald, P. S.; Spratt, P. M. Skin cancer in Australian heart transplant recipients. *J. Am. Acad. Dermatol.* **1999**, *40*, 27–34.
62. Final monograph for sunscreen drug products for over-the-counter human use. *Fed. Regist.* **1999**, *64* (98), 27666–27693.
63. Tuchinda, C.; Lim, H. W.; Osterwalder, U.; Rougier, A. Novel emerging sunscreen technologies. *Dermatol. Clin.* **2006**, *24*, 105–117.
64. Reisch, M. New-wave sunscreens active ingredient makers are frustrated by the long list of sunscreens and UV-A testing protocols that are still waiting FDA decisions.*Chem. Eng. News* **2005**, 18–22.
65. Magrez, A.; Horváth, L.; Smajda, R.; Salicio, V. r.; Pasquier, N.; Forró, L. S.; Schwaller, B. Cellular toxicity of TiO^2-based nanofilaments. *ACS Nano* **2009**, *3*, 2274–2280.
66. Dunford, R.; Salinaro, A.; Cai, L.; Serpone, N.; Horikoshi, S.; Hidaka, H.; Knowland, J. Chemical oxidation and DNA damage catalysed by inorganic sunscreen ingredients. *FEBS Lett.* **1997**, *418*, 87–90.
67. Gasparro, F. P.; Mitchnick, M.; Nash, J. F. A review of sunscreen safety and efficacy. *Photochem. Photobiol.* **1998**, *68*, 243–256.
68. Shaath, N. A. The chemistry of sunscreens. *Cosmet. Toiletries* **1986**, *101*, 55–70.
69. Diffey, B. L.; Tanner, P. R.; Matts, P. J.; Nash, J. F. In vitro assessment of the broad-spectrum ultraviolet protection of sunscreen products. *J. Am. Acad. Dermatol.* **2000**, *43*, 1024–1035.
70. Stevens, M. P. *Polymer Chemistry: An Introduction*, 3 ed.; Oxford University Press: New York, 1999.
71. Rahman, M.; Brazel, C. S. The plasticizer market: An assessment of traditional plasticizers and research trends to meet new challenges. *Prog. Polym. Sci.* **2004**, *29*, 1223–1248.

Chapter 2

Novel Thermoplastic Polyurethane Elastomers Based on Methyl-12-Hydroxy Stearate

Omprakash S. Yemul and Zoran S. Petrović*

Kansas Polymer Research Center, Pittsburg State University, 1701 S. Broadway, Pittsburg, KS 66762
***zpetrovi@pittstate.edu**

Novel vegetable oil-based elastomers were prepared from hydrogenated ricinoleic acid methyl esters. Polyester polyol with molecular weight 2500 was synthesized by transesterification of methyl-12-hydroxy stearate and 1,6-hexanediol. Segmented polyurethanes with 50 and 70% soft segment concentration (SSC) were prepared using the prepolymer method by reacting polyester polyol with diphenylmethane diisocyanate (MDI) and 1,4-butanediol as chain extender. Segmented polyurethanes were soluble in polar aprotic solvents such as DMF, DMAc, DMSO and NMP at room temperature. The polymers were characterized by GPC, viscometry and spectroscopic methods. Thermal and mechanical properties of polymers indicated good micro-phase separation. Both soft and hard segment displayed a certain degree of crystallization. While tensile strengths were in the expected range, elongations at break were low presumably due to lower molecular weights.

In recent years, the use of renewable bio-based materials has attracted much attention in scientific and industrial areas. This interest is catalyzed in part by the energy and environmental considerations. Oils of vegetable origin make up the greatest portion of current consumption of renewable materials in the chemical industry. They are sustainable, biodegradable, and offer to chemistry several possibilities of applications that can rarely be met by petroleum chemistry (*1*, *2*).

One of the major applications for vegetable oil based feedstock is in the polyurethane industry, especially in (rigid, flexible) foams, coatings, and elastomers which have been studied recently (*1–3*). Segmented polyurethanes are block copolymers with alternating soft and hard blocks which due to structural differences/incompatibilities, separate into domains formed from repetitive hard/soft segments. Polyurethanes with 70% soft segment concentration (SSC) are soft thermoplastic rubbers, whereas those with 50% SSC are hard rubbers, both types being of significant industrial importance (*4–7*). Segmented polyurethanes can be processed by injection molding and extrusion. Segmented polyurethanes have good properties and a wide range of applications such as sporting goods, environmentally responsive agents (*8*), internal release agents (*9*), controlled release of drugs (*10*), liquid crystalline polymers (*11*), and biomedical appliances (*12*) such as prosthetic tubes, arterial cannulas etc.

Methyl-12-hydroxy stearate (MHS) is an important industrial chemical derived from castor oil. It has been used in lubricants (*13*), and coatings formulations, and its polyesterpolyol has been used in making polyurethane foams (*14–16*). The advantage of methyl-12-hydroxy stearate over methyl ricinoleate is that it has no unsaturation, thereby eliminating any possible coloration to the final product. In our previous study (*6, 7*) we have shown that methyl ricinoleate based segmented polyurethanes with varying hard segments displayed micro-phase separation. In continuing this work on segmented polyurethanes in the present article, we report the synthesis of segmented polyurethanes with methyl-12-hydroxy stearate-based soft segment.

Experimental

Materials

Methyl-12-hydroxy stearate (98.9% pure) was a gift from Vertellus Performance Materials, Greensboro, NC, USA. MHS was recrystallized from methanol: water mixture, melting point 54 °C (lit (*17*) 55 °C). Titanium (IV) isopropoxide was obtained from Acros Organics USA. 1, 4-Butanediol was purified by vacuum distillation prior to use, 1,6-hexamethylenediol, anhydrous *N*, *N'*dimethylacetamide (DMAc), and *N,N'*-dimethylforamide (DMF) was purchased from Sigma-Aldrich and used as received. Mondur M from Bayer is pure diphenylmethane diisocyanate (MDI) with melting temperature 40°C. It was purified by vacuum distillation.

Synthesis of Polyesterpolyol

Polyesterpolyol was synthesized via polycondensation (transesterification) of methyl-12-hydroxy stearate with 1,6-hexanediol as initiator as follows: In a three necked round bottom flask (equipped with overhead stirrer, Dean Stark, nitrogen inlet and thermometer) was placed 53.46 g (0.17 mol) of methyl-12-hydroxystearate, 2.36 g (0.019 mol) 1,6-hexanediol, and 0.10 g titanium isopropoxide. The whole reaction mixture was heated while stirring under nitrogen bubbling. After 30 minutes the temperature was raised to 150°C and methanol was distilled out through the side arm. Methanol was removed from the side arm for measurement. After 60 minutes mild vacuum was applied for 45 minutes and finally the reaction was maintained at 190 to 200°C for 130 minutes. The reaction product was then cooled to room temperature and characterized. The reaction pathway is presented in Scheme 1. Yield 48 g; Viscous liquid; MW (2793); Viscosity@25°C 1.51 Pa.s; OH# number: 41.96; FTIR (Neat, KBr): 3535, 3451 cm^{-1} (-OH group); 2924, 2856 cm^{-1} (sym and asym $-CH_2-$ stretching band); 1731 cm^{-1} (C=O ester band stretching); 1461 cm^{-1} (C-O bending), 1174 cm^{-1} (-COO- stretching), 719 cm^{-1} (methylene stretching band). ^1H NMR ($CDCl_3$, TMS, 300 MHz): 4.07 ppm (t, 4H $-OCH_2$), 3.65 ppm (d, 2H - CH), 2.34 ppm (m, 4H $-CH_2$-CO), 1.60 – 1.25 ppm (m, 62H $-CH_2-$), 0.89 – 0.85 ppm (m, 6H $-CH_3$).

Synthesis of Segmented Polyurethanes

Thermoplastic segmented polyurethanes were synthesized by a one-pot two step polymerization process. A typical reaction is as follows: A prescribed amount of polyesterpolyol, (pre dried under vacuum at 75 – 80°C for about 2 hours), was transferred to a 100 mL three neck round bottom flask equipped with overhead stirrer, thermometer, nitrogen inlet and reflux condenser. Under nitrogen protection, a predetermined amount of MDI was added into the flask. The whole mixture was stirred under nitrogen atmosphere at 60 to 80°C for 3 h. After 3 h, the solution became viscous. Anhydrous DMAc was then added to the prepolymer at the amount required to make up 30 wt% of the final solution. After addition of a prescribed amount of 1,4-butanediol into the reaction mixture with a syringe, the temperature of the reaction mixture was raised to 90°C and the polymerization was continued overnight. The polyurethanes were purified by precipitating into water. The precipitated granules were separated by filtration and thoroughly washed with water, followed by vacuum drying to constant weight. Polyurethanes with 50% SSC (TPU50) and 70% SSC (TPU70) were prepared by this procedure. The general reaction pathway is presented in Scheme 2.

Preparation of Cast Films for Characterization

Segmented polyurethanes films were prepared by solution casting as follows: a 100 mL round bottom flask equipped with magnetic stirrer bar was charged with 12 g of polymer and 28 g of dry distilled DMF to obtain a 30 wt% solution. The flask was then heated on an oil bath at 85°C. The solution was cast onto a piece of freshly cleaned glass plate (10 cm X 10 cm) which was preheated in a force-draft oven at 85°C. The oven temperature was maintained at 80°C for 6 hours, after which the glass plate was taken out and cooled to room temperature and the film peeled out from glass plate by dipping into water. Finally, any residual solvent was removed by drying in a vacuum oven till constant weight.

TPU50 (50% SSC)

FTIR (ATR): 3329, 2923, 2850, 1701, 1596, 1523, 1413, 1311, 1222, 1064, 1015, 812, 751, 666 cm^{-1}; ^{1}H NMR (DMSO-d_6, 300 MHz, TMS, δ ppm): 9.50 (s, -COO-NH-), 7.32 (s, Ph), 7.06 (s, Ph), 4.08 (s, -CH_2-), 3.76 (s, -O-CH_2-), 2.09 – 0.77 (m broad, polyester). DSC: T_m -18°C (4.68 J/g), T_g 100°C, T_m 175°C (19.34 J/g);

TPU70 (70% SSC)

FTIR (ATR): 3323, 2915, 2848, 1731, 1701, 1596, 1528, 1461, 1410, 1376, 1309, 1216, 1056, 1014, 912, 807, 769, 719, 656 cm^{-1}; ^{1}H NMR (DMSO-d_6, 300 MHz, TMS, δ ppm): 9.50 (s, -COO-NH-), 7.32 (s, Ph), 7.06 (s, Ph), 4.08 (s, -CH_2-), 3.76 (s, -O-CH_2-), 2.09 – 0.77 (m broad, polyester). DSC: T_m -14°C (6.65 J/g), T_g 100°C, Tm 180°C (7.61 J/g);

Methods

The hydroxyl values were determined according to ASTM E 1899-97 using the reaction with p-toluenesulfonyl isocyanate (TSI) and potentiometric titration of the resulting carbamate with tetrabutylammonium hydroxide. The polyol viscosity was measured on the Rheometrics Scientific Inc. (Piscataway, NJ) SR- 500 dynamic stress rheometer in a stress-controlled mode, between parallel plates of 25 mm in diameter and a gap of 0.20 mm. IR spectra were recorded on a Perkin–Elmer (Waltham, MA) Spectrum-1000 Fourier transform infrared (FT-IR) spectrometer. ^{1}H NMR spectra were recorded using a Bruker AV-500 spectrometer (Bruker, Rheinstetten, Germany) in deuterated chloroform/DMSO. The gel permeation chromatograms (GPC) were acquired on a Waters system consisting of a 510 pump and 410 differential refractometer. Flow rate of tetrahydrofuran eluent was 1 mL/min at room temperature. Calibration was performed using nine polystyrene standards. A differential scanning calorimeter, (model Q100, from TA Instruments, New Castle, DE, USA), was used to measure glass transition and melting points at a heating rate of 10°C/min. Dynamic mechanical tests were carried out on a DMA 2980 from TA Instruments at 1 Hz and heating rate of 5°C/min. Tensile properties were measured according to ASTM D882-97 using a tensile tester model 4467 from Instron, Canton, MA, USA.

Results and Discussion

The polyesterpolyol prepared by the transesterification reaction between methyl-12-hydroxy stearate and 1,6-hexanediol reaction is presented in Scheme 1. This polyesterpolyol was further used in the preparation of segmented polyurethanes. The reaction pathway is depicted in Scheme 2.

Structure of Polyesterpolyol and Segmented Polyurethanes

The FTIR spectrum of the polyesterpolyol shows a hydroxyl group stretching broad band at 3535 cm^{-1}, the hydroxyl peak in the hydrophobic environment splits into two bands at 3535 and 3451 cm^{-1}, the former being assigned to the non-hydrogen bonded and the latter to hydrogen bonded hydroxyl. It also shows bands at 2924, 2856 cm^{-1} corresponding to sym and asym $-CH_2-$ stretching respectively. Polyesterpolyol is characterized by a free ester carbonyl stretching band at 1731 cm^{-1}, and an ester C-O stretching band at 1461 cm^{-1}. The structure was confirmed by the proton NMR spectrum. It shows proton signals (all values in ppm) at 4.07 for $-OCH_2$, 3.65 for -CH, 2.34 for $-CH_2-CO$, 1.60 – 1.25 multiplet for methylene protons, and 0.89 – 0.85 for terminal methyl proton. There is no signal at 5.33 ppm indicating no double bond formation in the polyesterpolyol. The FTIR spectrum of segmented polyurethane shows characteristic urethane bands 3329 cm^{-1} for N-H stretching bond, free carbonyl band at 1733 cm^{-1} and amide II absorption at 1523 cm^{-1}. It also shows bands at 2923 and 2850 cm^{-1} symmetric and asymmetric for $-CH_2-$ stretching band. Proton NMR signals (values in ppm) at 9.50 for urethane -COO-NH-, 7.32 – 7.06 for aromatic protons from MDI, 4.08 – 3.76 for methylene protons of MDI, and 2.09 – 0.77 for polyester backbone.

Molecular Weight Distribution of Polyesterpolyol and Segmented Polyurethanes

The GPC data for the polyesterpolyol show the molecular weight distribution typical for polycondensation polymers. Low molecular weight species are observed as separate peaks while high molecular weight species merged into a single peak, Figure 1. The TSI method gave an OH# 41.9 (average of three measurements) from which the number average molecular weight of polyesterpolyol is calculated to be 2793 g/mol. M_n from GPC is 2509, the experimental molecular weight value is higher (i.e. 2793 from OH#) than the target molecular weight of M_n 2500 based on the assumption that all 1,6-hexanediol reacted with methyl-12-hydroxy stearate. At complete conversion of methyl ester only a polyester diol is formed, while at lower conversion a monol is formed together with the diol. The higher molecular weight of the diol is due to loss of 1,6-hexanediol during the vacuum process at high temperature. A similar observation was reported in previous publications (*6*, *7*, *18*). GPC and viscosity data of polyesterpolyol, polyurethanes are presented in Table I. The results show that an increase in soft segment content led to polyurethanes with lower molecular weight. The number average molecular weights relative to polystyrene were 27,050 (PDI 4.7) for TPU50, 19680 (PDI 1.5) for TPU70. Intrinsic viscosities of samples measured in DMF are (0.42 for TPU50 and 0.28 for TPU70) below 0.5 dL/g, indicating the relatively low molecular weights which would likely affect the mechanical properties.

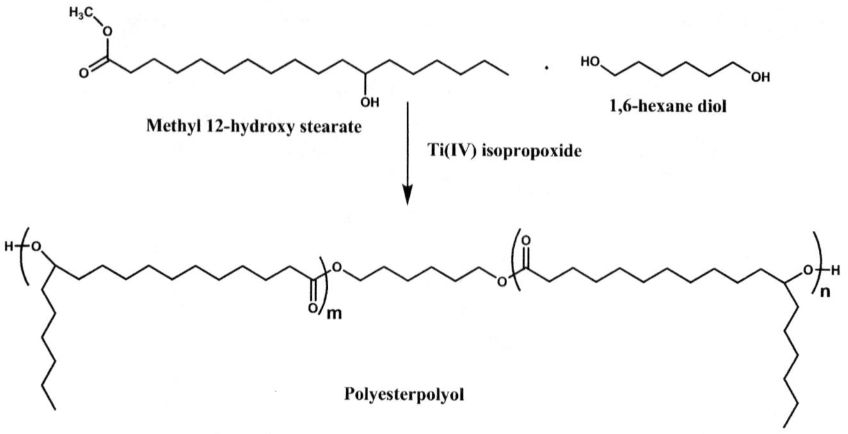

Methyl 12-hydroxy stearate

1,6-hexane diol

Ti(IV) isopropoxide

Polyesterpolyol

Scheme 1. Synthesis of Polyesterpolyol by transesterification of Methyl 12-hydroxy stearates and 1,6-hexanediol catalyzed by titanium isopropoxide

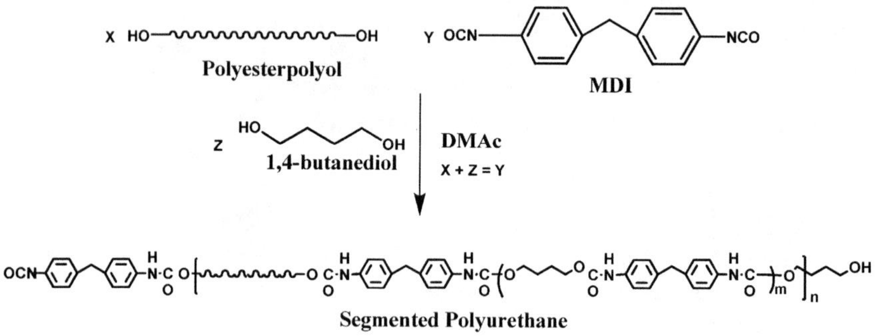

Polyesterpolyol

MDI

1,4-butanediol

DMAc

X + Z = Y

Segmented Polyurethane

Scheme 2. Synthesis of Segmented Polyurethanes by solution polymerization

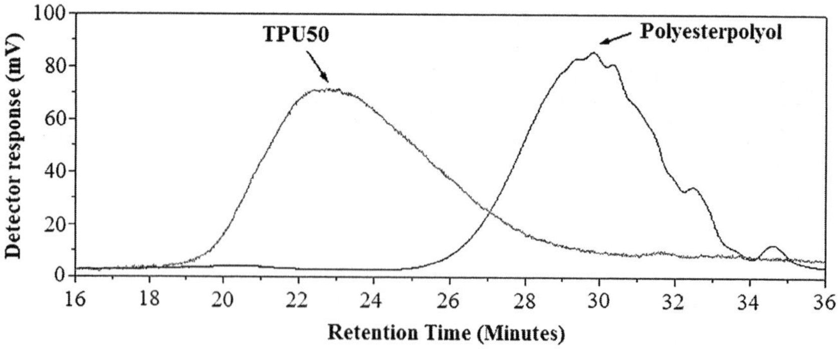

Figure 1. GPC traces of the polyesterpolyol and segmented polyurethane (TPU 50)

Table I. GPC and Viscosity Data of Polyesterpolyol and Segmented Polyurethanes (TPU50 50% SSC) and (TPU70 70% SSC)

Sample	*Mn*[a]	*Mw*[a]	*PDI*[a]	*Intrinsic Viscosity*[b]
Polyesterpolyol	1552	2509	1.61	-
TPU50 (50% SSC)	27053	128954	4.79	0.42
TPU70 (70% SSC)	19682	30230	1.51	0.28

[a] Data from GPC (using polystyrene standard. [b] In DMF @25°C.

Thermal Behavior of Segmented Polyurethanes

The DSC scan of polyesterpolyol shows a broad melt transition at -1.5°C when heated from -90°C to 100°C with melt enthalpy of 68.6 J/g. However, no glass transition is observed for the polyesterpolyol. Similar compounds based on polyricinoleate show glass transition at -76°C (*6*, *7*). The melting of polyesterpolyol could be due to the long chain saturated fatty acid chains' crystallization (*19*). Segmented polyurethanes are typically characterized by three transitions related to a glass transition of the soft segment, a glass transition of amorphous phase of the hard segment and a melting of the hard phase. The segmented polyurethanes of the present study display melting and glass transition for both soft and hard segments. The melting of the soft segment at -18 °C could be due to the melting of the saturated long chain fatty chains and the melting of the hard segment is due to the melting of the MDI and butane diol segments. Both polymers show glass transition at 100 °C related to the hard segments, which is similar to previously reported polyricinoleate based segmented polyurethanes (*6*, *7*). The glass and melt transition temperature values of the polymers are listed in Table II.

TPU50 has melt enthalpy of 19.3 J/g, which is closer to polyurethanes (i.e. 25 J/g) with 50% polypropylene oxide soft segment. However melt enthalpy of TPU70 with 70% soft segment is 7.6 J/g vs. 5 J/g for the PPO based analogues (*5*). A clear melting point (Figure 2) (170 - 190°C) in polyurethanes based on methyl-12-hydroxy stearate with TPU70 (70% SSC) indicates better phase separation. Comparative DSC scans of polyesterpolyol, TPU50, and TPU70 are presented in Figure 2.

Dynamic Mechanical Properties of Segmented Polyurethanes

Dynamic mechanical analysis (DMA) tests were carried out to obtain further information on the micro phase separation. TPU50 and TPU70 display the dependence of storage and loss modulus on temperature. TPU50 (50% SSC) shows a higher rubbery plateau modulus than TPU70 (70% SSC), which is due to a higher SSC in TPU50 than TPU70. The modulus of TPU50 that contained only 50% SSC remained high until 100°C while the modulus of TPU70 with 70% SSC showed a dramatic decrease of several order of magnitude above this temperature. The DMA plots of TPU50 and TPU70 are presented in Figure 3. The glass transition temperature values estimated from the loss modulus were

Table II. Glass transition and Melt temperature of Segmented Polyurethanes from DSC scans (values in bracket indicate melt enthalpy)

Sample	*Soft SSC*[a]		*Hard SSC*[a]	
	Tg(°C)	*T_m(°C)/J/g*	*T_g(°C)*	*T_m(°C)/J/g*
TPU50	-39	-18 (4.7)	100	175 (19.3)
TPU70	-40	-14 (6.6)	100	180 (7.6)

[a] Data obtained from DSC scans heating rate@10 °C/min under nitrogen atmosphere.

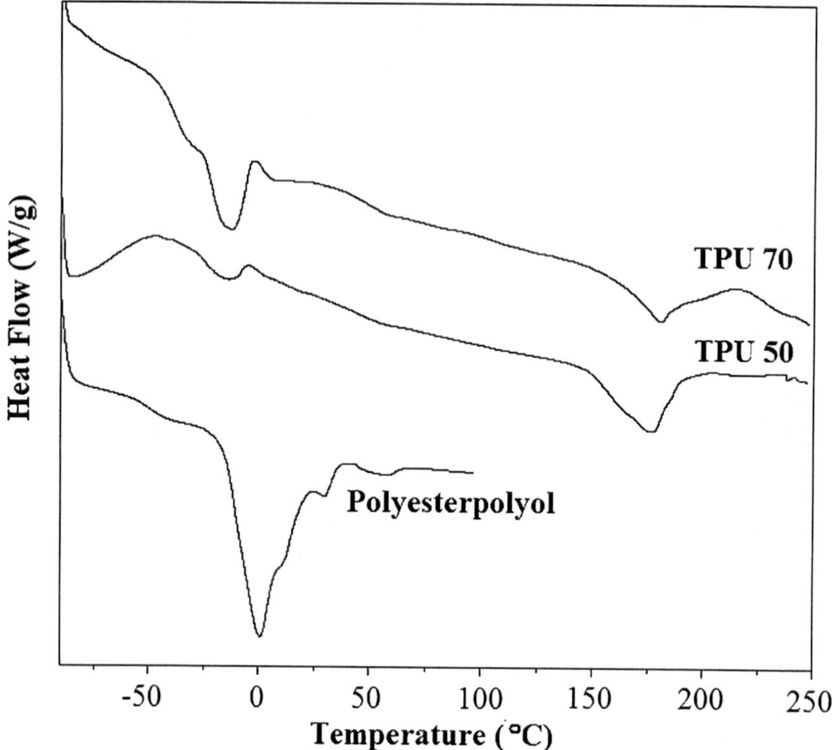

Figure 2. DSC scans of polyesterpolyol, TPU50 and TPU70

virtually the same, being for TPU50 (-27°C) and for TPU70 (-26°C), respectively. Invariability of T_g with SSC content indicates good phase separation. The co-continuous morphology of the TPU50 is characterized by a small drop of storage modulus above T_g of the soft segment, while this drop is 2-3 orders of magnitude for the dispersed hard domains in a soft segment matrix in the TPU70. The maxima on the loss modulus curves characterizing the α-transition did not vary between the samples, indicating a good phase separation.

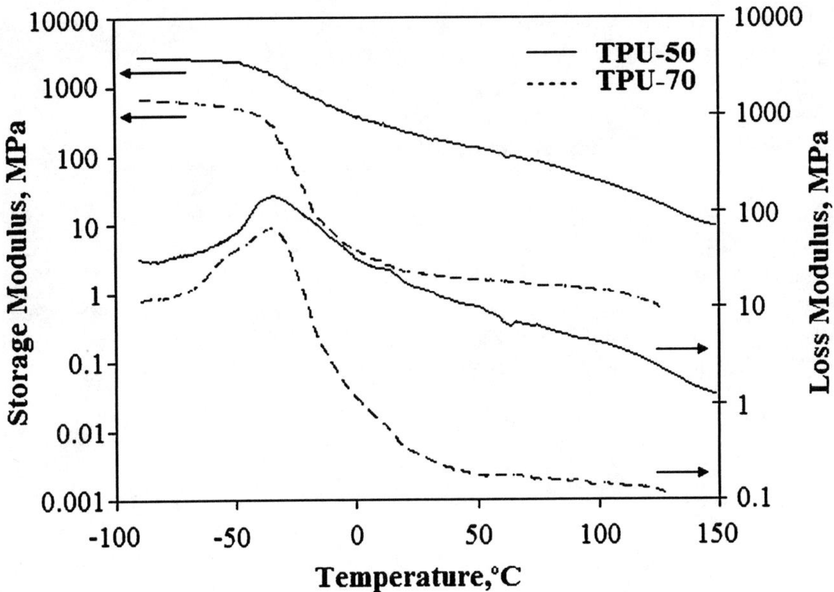

Figure 3. Effect of temperature on storage and loss modulus for TPU50 (solid) and TPU70 (dashed line).

Tensile Properties of Segmented Polyurethanes

Mechanical properties of segmented polyurethanes are closely related to their composition. Stress-strain behavior of polymers, which were uniaxially deformed until failure, showed that TPU50 had higher modulus than TPU70. Modulus values calculated at 50% elongation for TPU50 were 310 kPa while TPU70 had 40 kPa. TPU50 has tensile strength of 19 MPa and elongation break at 134% while these values decreased to 2.4 MPa and 55% elongation for TPU70. The difference in stress-strain behavior of TPU50 (50% SSC) and TPU70 (70% SSC) arises from the difference in continuity of hard and soft micro phases. Poor elastic properties of segmented polyurethanes with 70% SSC were ascribed to the low molecular weight of the polymer as well as to the presence of the higher content of six carbon dangling chains, which have a stronger effect on strength than in the sample with co-continuous morphology.

Thermal Stability of Segmented Polyurethanes

The thermal stability of the polyurethanes was measured by TGA. Polyurethanes generally possess low thermal stability due to the presence of urethane bonds. Both polymers are thermally stable up to 300°C. The TGA and derivative TGA (DTGA) curves under nitrogen shows two stage decomposition with rate of decomposition maxima at 330°C for TPU50 and 360 °C for TPU70. Comparative TGA curves of TPU50 and TPU70 are presented in Figure 4. The decomposition in the region 400 - 500 °C correlates well with the aromatic content of the polymer. The onset of degradation is about 300 °C, i.e., high enough to allow safe processing by extrusion and injection molding.

Optical Properties of Polyurethane Films

Films made from both polymers are yellowish-transparent as shown in Figure 5. Usually TPU's with crystallizing hard segments are not transparent at 50% SSC due to scattering of light on the large hard domains. Transparency at 50% SSC in our case indicates either that crystallites or phases are smaller than the wavelength of light or that the refractive index of the phases is similar. It appears that the small hard domain size is the determining factor in our case.

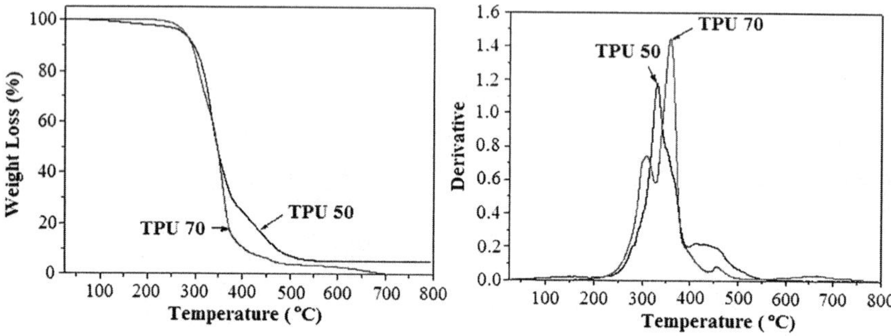

Figure 4. Comparative TGA and DTGA scans of TPU50 and TPU70 under nitrogen atmosphere.

Figure 5. Illustration of the film transparency- left: PU50 SSC, right: PU70SSC.

Conclusions

New bio-based segmented polyurethanes with 50% soft segment and 70% soft segment were synthesized from polyesterpolyol based methyl-12-hydroxy stearate. The solution cast films of both TPU50 and TPU70 are transparent. Thermal studies show that segmented polyurethanes are thermally stable enough to be processed by injection molding and extrusion. These materials have potential applications in drug delivery and biomedical devices.

Acknowledgments

This work was supported by research funding from U.S. Department of Agriculture, Award No. 2008-38924-19200.

References

1. Lubick, N. *Environ. Sci. Technol.* **2007**, *41*, 6639.
2. Petrović, Z. S. *Polym. Rev.* **2008**, *48*, 109.
3. Metzger, J. O.; Bornscheuer, U. *Appl. Microbiol. Biotechnol.* **2006**, *71*, 13.
4. Petrović, Z. S.; Fergusan, J. *Prog. Polym. Sci.* **1991**, *16*, 695.
5. Petrović, Z. S.; Javni, I.; Divjakovic, V. *J. Polym. Sci., Part B: Polym. Phys.* **1998**, *36*, 221.
6. Petrović, Z. S.; Xu, Y.; Zhang, W. *Polym. Prepr.* **2007**, *48*, 852.
7. Xu, Y.; Petrović, Z. S.; Das, S.; Wilkes, G. *Polymer* **2008**, *49*, 4248.
8. Vaidya, A.; Chaudhury, M. K. *J. Colloid Interface Sci.* **2002**, *249*, 235.
9. Uchida Y.; Yoshida Y.; Kaneda T.; Moriya T.; Kumazawa T. U.S. Patent 5,306,788, 1994.
10. Batyrbekov, E. O.; Iskakov, R. M.; Boldyrev, D. Y.; Yu, V. K.; Praliev, K. D.; Zhubanov, B. A. *Mater. Res. Soc. Symp. Proc.* **2004**, *EXS-1*, 433.
11. Tang, W.; Farris, R. J.; MacKnight, W. J.; Eisenbach, C. D. *Macromolecules* **1994**, *27*, 2814.
12. Boretos, J. W.; Pierce, W. S. *Science* **1967**, *158*, 1481.
13. Teeter, H. M.; Gast, L. E.; Bell, E. W.; Cowan, J. C. *Ind. Eng. Chem.* **1953**, *45*, 1777.
14. Peerman, D. E.; Rogier, E. R. U.S. Patent 4,423,162, 1983.
15. Van Der Wal, H. R.; Bartelink, C. F. International Patent WO2006/047433, 2006.
16. Stutts, K.; Prange, R.; Zhang, M.; Schrock, A. K. International Patent WO2006008136, 2006.
17. Walter, J. G. *J. Am. Chem. Soc.* **1958**, *80*, 4593.
18. Petrović, Z. S.; Cvetković, I.; Hong, D.; Wan, X.; Zhang, W.; Abraham, T.; Malsam, J. *J. Appl. Polym. Sci* **2008**, *108*, 1184.
19. Dunn, R. O. *J. Am. Oil Chem. Soc.* **2008**, *85*, 961.

Chapter 3

Natural Fibers Production, Processing, and Application: Inventory and Future Prospects

Ryszard M. Kozlowski,[*,1] Maria Mackiewicz-Talarczyk,[2] and Jorge Barriga-Bedoya[2]

[1]The Institute for Engineering of Polymers Materials and Dyes, Harcerska str.30, 05-820 Piastow, Poland
[2]The Institute of Natural Fibres & Medicinal Plants, Wojska Polskiego str. 71b, 60-630 Poznan, Poland
[*]E-mail: ryszard.kozlowski@escorena.net

The chapter presents the inventory, achievements, and future prospects in the area of natural fibers production, processing and application, with special emphasis on the fibrous plants.Various parts of lignocellulosic plants like seeds, leaf, straw, bast, woody core have potential as the sources of valuable textile fibers, and are also suitable for use in automotive and aerospace industries, building materials, and biopolymers. Production of natural fibers does not damage the ecosystem. Plants, which are the source of natural fibers recycle the carbon dioxide. Fibrous plants remediate soil polluted by heavy metals. Nowadays the position of natural fibers in the world fiber market and the level of production is stable, thanks to the growing area of their application, not only in textiles (woven, knitting, non-woven, technical textiles), but in more eco-friendly composites, for production of agro-fine-chemicals, medicines and cosmetics as well as for nourishment.

High biomass of some of fibrous plants can be used as a source of pulp and energy (solid fuel) and for the production of bio-ethanol and n-buthanol. Also GM modification brings new possibilities towards higher productivity, better quality and greater area of application of natural raw material.

Introduction

Nature in its abundance offers us a lot of material than can be called fibrous. Fibers can be found in plant leaves, fruits, seed's cover or stalk. Plants of that structure grow under any geographical width we can name. They all have one common characteristic-all are environmentally friendly in 100%. We can talk about them as totally renewable and biodegradable.

All fiber types suffered from slowing demand in 2008/2009. Manmade fibers, strongly established, like polyester, polyamide, polypropylene and acrylic were down in volumes and decreased by 4.5% to 42.2 million tonnes. In 2009 the global share of manmade fibers in fibers market increased to 63%. The usage of cotton, wool and silk also decreased by 10.1% to 25.2 million tonnes. Bast and hard fibers such as jute, ramie, flax, hemp, kapok, sisal and coir is anticipated to have stagnated at 5.9 million tonnes. Small-scale fiber types like aramid and carbon fibers weathered the downturn not bad until the fourth quarter of 2008. Although firm demand was observed in aerospace, automotive, military and wind power and it created the positive growth rate (*1*).

The Market of Fibers and Fiber Based Products

Global fibers consumption 1980-2008 and forecast for 2009/2010 is presented on figure 1.

Up to the year 2008 we observe (six year) growth of consumption (see table I), in 2008 year consumption dropped 6.7% at 67.3 millions tonnes and in 2009 reached 70.5 millions tonnes (*1*).

Global (2004-2008) fiber production is presented on figure 2.

The world inventory of fibers was estimated befor 2009 at 74.26 million tonnes: where share of cotton 25.0; wool 1.16; other natural fibers 5.9, manmade fibers - PES, PA, PP, PAN: 42.2 [millions of tonnes] (*2*). The global fiber demand in 2009 achieved 70.5 millions tonnes; manmade fibers- 44.1, natural fibers advanced to 26.4 millions tonnes, where cotton 22.3, wool 1.1, other natural fibers 3.0 [millions of tonnes] (*1*)

World nonwoven production between the years 2008 to 2010 is given on figure 3, and production of nonwoven by technology is on figure 4.

Emerging and dynamically growing fibers application for aircraft between 1990-2008 is presented on fig. 5.

The main area of application of fibers in aircraft industry is the usage of them in upholstery inside aircrafts, carbon fibers as reinforcing composites, especially in the newest planes.

Lignocellulosic Fibers - Classification, Availability and Properties

The classification of lignocellulosic fibers is presented in the table II. The fibers are present in the diversified parts of the plants such as: seed, bast, leaf, and straw. Wood, cane, and grass are also the source of the fibers.

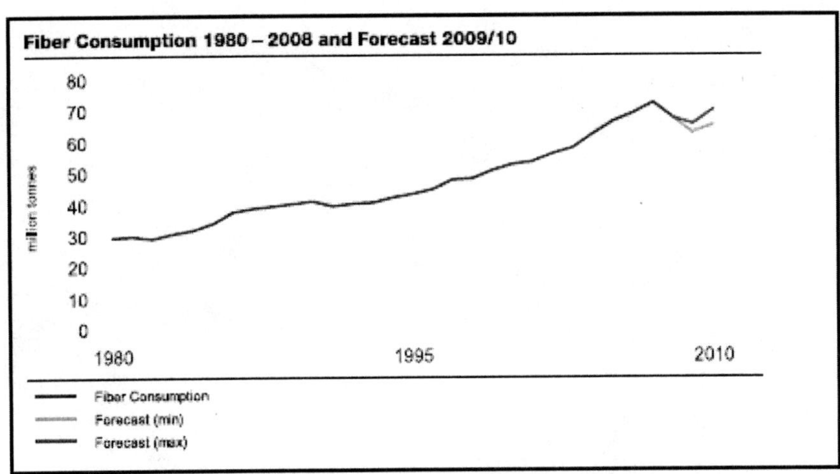

Figure 1. Fiber consumption 1980-2008 and forecast 2009/10. Reproduced with permission from reference (1). Copyright 2009 Oerlikon

Table I. Consumption of natural fibers (*1*). *Reproduced with permission from reference (1). Copyright 2009 Oerlikon*

Year	*Cotton*	*Wool*	*Silk*	*Total*	*+- in %*
2008	23,907	1,164	100	25,171	-10.2
2007	26,713	1,201	98	28,012	-0.7
2006	26,875	1,232	98	28,205	5.9
2005	25,328	1,216	97	26,641	6.7
2004	23,658	1,214	98	24,970	10.1

"Green" plant fibers like flax, cotton, jute, sisal, kenaf and fibers of allied plants, which have been used since more than 6000 years BC presently are and will be the future raw materials not only for the textile industry as well for: modern eco-friendly composites used in different areas of application like building materials, particleboards, insulation boards, food, feed and nourishment, friendly cosmetics, medicine and sources for other bio-polymers and "agro fine chemicals" (*6, 7*).

They do not cause any disturbing effects of ecosystem, they can grow in different climatic zones and they recycle the carbon dioxide for the atmosphere of our Earth (*8*). "Resurrection" of these plants for world is very important, because they provide better agriculture balance in the World and they will reduce a deficit of cellulosic pulp for this Millennium when the population will be multiplied up to about 11.6 billion people. Some of these green plants like flax and hemp, kenaf also can be used for cleaning soil, polluted by heavy metals, be extracting and removing cadmium (Cd), lead (Pb) and copper (Cu) and others (*9–11*).

Nowadays there is a "green light" for expanded use of green fibers in textiles for healthy, comfortable clothing, which after lifetime- will be fully recyclable/ biodegradable. The future demand for comfort providing hydrophilic fibers is

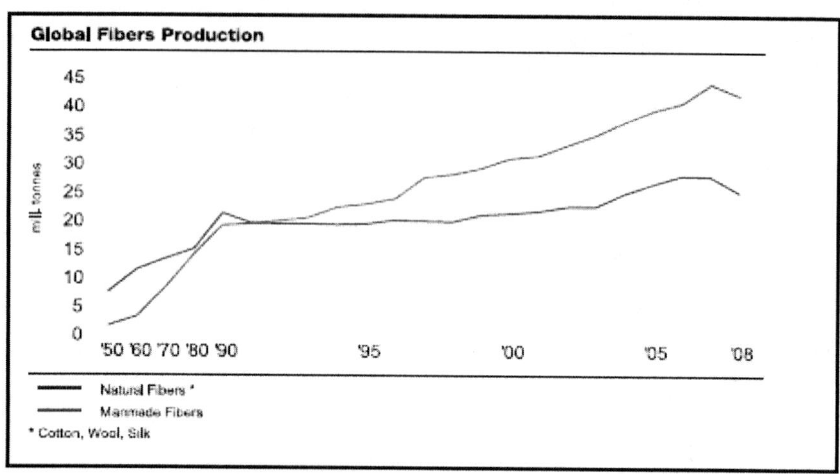

Figure 2. Global fibers production. Reproduced with permission from reference (1). Copyright 2009 Oerlikon

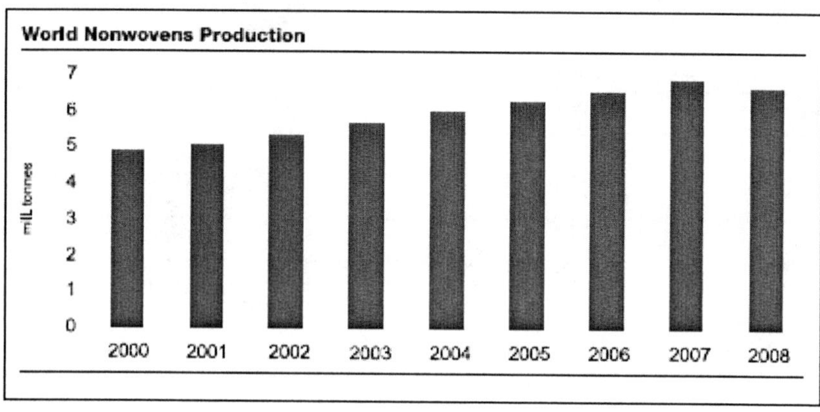

Figure 3. World nonwovens production. Reproduced with permission from reference (1). Copyright 2009 Oerlikon

predicted to be growing, namely for cotton to maintain at the level of about 22 millions tonnes up to the year 2050, while the growth of the demand for viscose, lyocell and other natural cellulosic fibers is expected at the following leves: about 3.0 in 2010; 5.0 in 2010; 10.0 in 2030; 15.0 in 2040 and about 17.5 in the year 2050 [millions tones] (*12*)

Fibrous plant raw materials are offered by NATURE like flax, hemp, kenaf, ramie, jute, sisal, abaca, curaua, coir, cabuya, pineapple and many others can be used both in textiles and composites solving difficult economical and environmental problems (*13–15*).

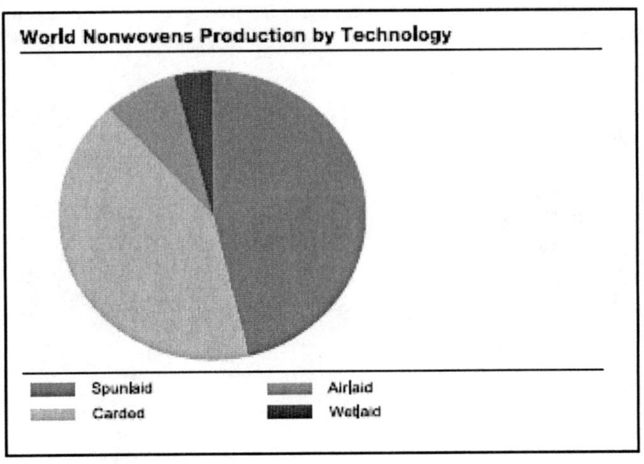

Figure 4. World nonwovens production by technology. Reproduced with permission from reference (1). Copyright 2009 Oerlikon

Main Directions in Application of Natural Fibers

TEXTILES: Woven, Nonwoven, Knitting, Technical, PULP, COMPOSITES, FINE CHEMICALS, ENERGY.

Special treatment likes enzymatic, liquid ammonia, plasma and corona treatment provides new promising features and properties of fibers like "fineness" and crease resistance (*4*, *16*). Concluding, in textile area of application these green fibers now are better adopted to modern processing techniques, which develop new clothing/home furnishing items, biodegradable nonwoven and geo-textiles. They adapt more aggressive marketing of their perfect hygiene/health properties and gain proper attention of governments, local authorities and international bodies and standards creators. According to newest research contact of body with natural, especially linen increase the alfa-immunoglobulin content in human organism and this immunoglobulin is a decisive factor in our immuno-resistance system. (*13*, *17*, *18*). What we also do know is that natural fibers involve lower miographic tension of muscles than synthetic fibers (*18*). Our last tests show that phenomena of allergy resulting from higher level of histamine in human blood is also an effect of man being exposed to synthetic fibers (*18*).

Efforts are taken also, under Food and Agriculture Organization of the United Nations (FAO) and EMBRAPA (Brazilian Agricultural Research Corporation's) patronage, to introduce the production of knitted apparels with curaua fiber within a welfare programme for Brazilian rural areas (*10*, *15*).

Natural fibers are characterized by such properties as: air permeability, hygroscopicity and capability of giving up moisture, no allergic effect (except for wool), biodegrability (*5*). Due to these properties natural fibers (including those unsuitable for spinning) become a very good material for stuffing and filling of upholstered furniture, particularly in the case of above mentioned products (*4*).

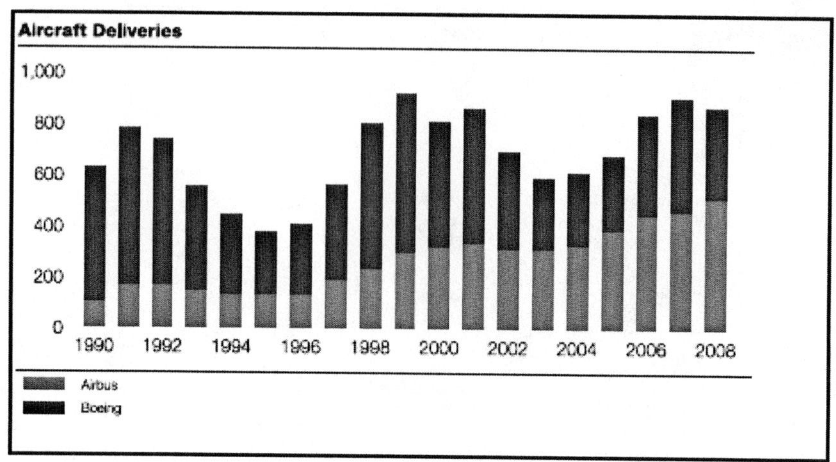

Figure 5. Aircraft deliveries. Reproduced with permission from reference (1). Copyright 2009 Oerlikon

Table II. Types of lignocellulosic fibers[a]

Type of fiber						
Wood	***Seed***	***Bast***	***Leaf***	***Cane***	***Straw***	***Grass***
Conifer-ous	Cot-ton	Flax	(Manila)	Bagasse	Corn	Esparto
Decidu-ous	Kapok	Hemp	Date-palm	Bamboo	Oat	Elephant grass
	Coir	Jute	Henequen	Reed	Linseed	Vetiver
		Kenaf	Pineapple	Banana	Rape	Reed
		Ramie	Sisal		Rice	Canary grass
		Roselle Karkadeh	Curaua		Rye	
			Cabuya		Wheat	
					Buckwheat	

[a] (*3–5*).

"Green" Plant Fibers/Bast Fibrous Plants for Pulp

In the cellulose industry the interest is growing in production of pulp and paper from agro based lignocellulosic raw materials such as bagasse, bamboo, reeds, esparto, hemp, flax, kenaf, abaca, sisal, grass, etc. The growing contribution of cellulose of agro plant origin to the world production of pulp and paper is noticed (*19*).

World population, forest area and wood consumption are presented on fig.6

In addition to cellulose, the main components of lignocellulosics are lignin, hemicelluloses, and pentosans. The role of isolated lignin in the chemical industry

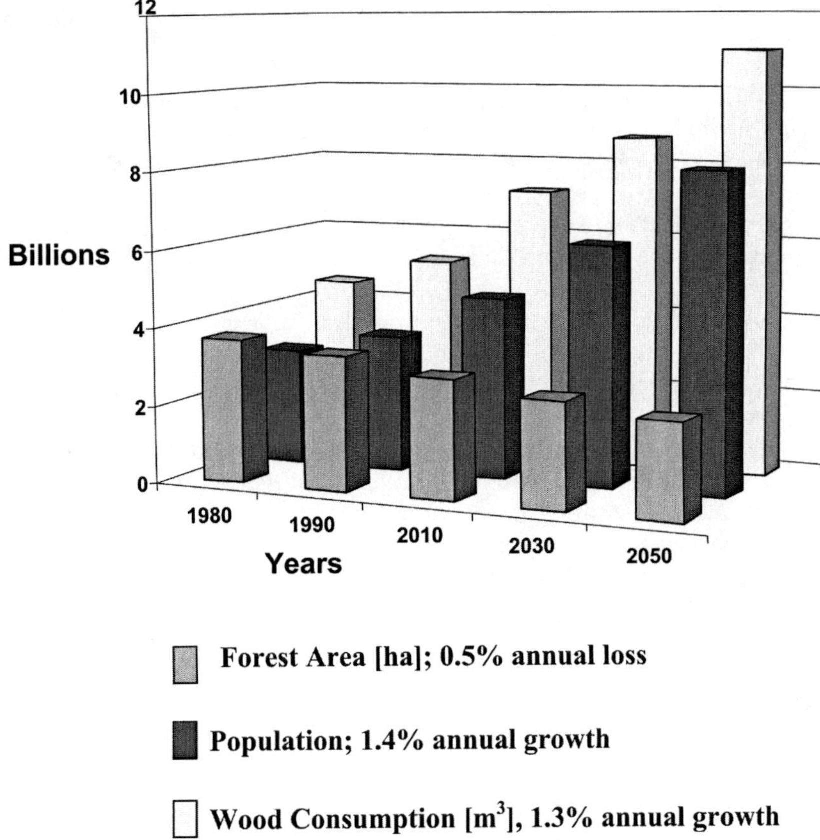

Figure 6. World population, forest area and wood consumption, years 1980-2050 (INF calculations)

grows incessantly. This by-product obtained during the manufacture of cellulose found its application, among others, as: substitute for phenol-formaldehyde resins in composites, a natural polymer which can be modified to higher reactivity, and applied as follows: agrochemical, for packaging, laminates, moisture barrier coatings, stiffening agent (boxboards), friction materials (brakes, pads), wood adhesives (plywood, wafer boards, particleboards, fireboards), plastic moulding (automotive), foundry mould binders, antioxidants and UV blockers (*4*, *20*). The application of nonwovens is very wide ranging from furniture to the geo-and chemo textiles. The latter are used for insulation and reinforcement of earth structures like landfills, slopes and embankments, grass mats etc. The surface of the newly formed slope can be mulched with bark, cut straw, sawdust or their mixture with peat, and the mulched surface is covered with a fine network until it becomes naturally strong (*5*).

Plant Fibers/Bast Fibrous Plants for Biocomposites, Based on Lignocellulosic Fibers

Biocomposites include a wide range of products for different applications ranging from construction or insulation panels made of wood pieces, particles and fibers, through special textiles (geo-textiles and nonwoven textiles), to plastic products based on polymers filled with lignocellulosic particles. In biocomposites: fiber plants: flax, hemp, ramie, kenaf, jute, abaca, sisal, coir, curaua are considered as promising lignocellulosic raw materials for different applications (*10*).

Composite materials find their application in car, aircraft (aviation), railway and truck industry. The population growth and economical development is accompanied by the development of motorization. The great advantage of composites reinforced with fibers is that when the fibers are arranged parallel to directions of applied forces (unidirectional laminates), the possibility of usage of anisotropic properties of material for structure arises (crushed fibers, embroided and 3-D weaved structures). The properties of this new material crashed, can be compared with regular glass fiber or reinforced polymers in normal conditions, for instance in automotive industry, road construction, irrigation system, landfills, furniture industry and also in sport. Broader field of polymer application causes the necessity to comply with stricter and stricter requirements for properties including the most important: mechanical strength, resilience and creeping, flammability, resistance to corrosive and weather conditions. There are several process techniques applicable for manufacturing composite materials. One of the best one is pultrusion which is a continuous process offering uniform cross sections of fiber reinforced profiles. They can be in form of rod, tube, angle, channel, and bar in more complex shape. The manufacturing process can be compared with conventional extrusion (*5*).

New Ecological Polymers from Vegetable Oils

In polymer applications derivatives of vegetable oils such as epoxides, polyols and dimerization products based on unsaturated fatty acids, are used as plastic additives or components for composites or polyurethanes.

One of the methods to obtain polymers from renewable resources is catalytic reaction of e.g. linseed oil with urea. It was found that in special conditions the linseed oil reacts quantitatively with urea

$$\begin{gathered}\mathrm{ROCOCH_2CH(OCOR)CH_2OCOR + 2\ NH_2CONH_2 \xrightarrow{cat,Temp}} \\ \mathrm{NH_2OCOCH_2CH(OCOR)CH_2OCONH_2 + 2\ RCONH_2}\end{gathered}$$

or indirectly, in a two-stage reaction, in homogeneous system

$$\mathrm{C_4H_9OH + NH_2CONH_2 \xrightarrow{cat,Temp} C_4H_9OCONH_2 + NH_3}$$

$$\begin{gathered}\mathrm{ROCOCH_2CH(OCOR)CH_2OCOR + 2\ C_4H_9OCONH_2 \xrightarrow{cat,Temp.}} \\ \mathrm{NH_2OCOCH_2CH(OCOR)CH_2OCONH_2 + 2\ RCOOC_4H_9\ ).}\end{gathered}$$

These studies allow for obtaining products capable of modifying properties of resins, changing for instance hydrophobic properties, elasticity, solubility, etc. Reactions involving the alkyl chain or double bond represent less than 10% found mainly in polymer applications of epoxides, polyols, and dimerization products based on unsaturated fatty acids. These are used as plastic additives or components for composites or polymers like polyurethanes. Polyols based on vegetable oils can be used for obtaining polyurethane materials including elastomers and foams with the cell structure and properties required for various applications (*21*, *22*).

Bast Fibrous Plants as a Source of Food, Fodder, Pharmaceutical Products, Cosmetics

Flax and hemp seeds are perfect raw materials for agriculturally based industries such as the production and processing of food, of pharmaceuticals and cosmetics, or of paint varnishes. Linseed-based food additives also contain lignans, which seem to be capable of slowing the cell division of some tumors. They are said to diminish or delay the onset of colon and breast cancer. Lignans also improve urinary function, helping prevent inflammation of the kidneys. Dietary studies has shown that four weeks on flaxseed enriched bread is able to lower cholesterol levels 7% to 9% (depending on the particular cholesterol fraction tested). Linseed proteins and mucilage are used in food products such as ice cream, powdered sauces and soups to improve smoothness and viscosity (*5*).

Bast Fibrous Plants for Agro-Fine-Chemicals

Latest discovery showed that some of fibrous plants are rich sources not only phytoestrogens (lignans), but cyclopeptides and waxes. In addition to cellulose, the main components of lignocellulosics are lignin, hemicellulose and pentosans. Obtaining these substances from natural renewable resources is important for later added value use in cosmetic and pharmaceutical industry. The uses of isolated lignin in the chemical industry grow continuously (*5*).

Conclusions

Global trends towards sustainable development have brought to light natural, renewable, biodegradable raw materials, among them bast fibers. Science and technology continue in extending their use in textile and other industries. Recent achievements and new applications of plant fibers and derived, associated products form the background for the following conclusions: fast growing population, eco- and health awareness create large space for future expansion of other than cotton natural cellulosic fibers. Present achievements in breeding, production, processing extended the use of natural fibers in textiles. To make the way for these, difficult in processing fibers, into textile products being beyond their reach for centuries, it was necessary to develop: novel, refined natural fiber types, adapted to modern spinning systems; novel, softer, finer 100% natural or blended with other natural and manmade yarns, among them knitted yarns; new crease resistant finishing treatments; novel products, which could meet the needs of not only demanding apparel sector. Also we observe the growing area of application of natural fibers in composite reinforced by 3-D fiber structures.

References

1. *The Fiber Year 2008/09 and 2009/10: A World Survey on Textile and Nonwovens Industry*; Issues 9 and 10, Oerlikon: Pfäffikon, Switzerland, May 2009/May 2010.
2. Saurer Report. *The Fiber Year 2007/2008*; Oerlikon: Pfäffikon, Switzerland, 2008.
3. Rowell, R. M.; Harrison, S. M. Proceedings of the International Kenaf Conference, March 1991.
4. *Textiles for Sustainable Development*; Ananjiwala, R., Hunter, S., Kozlowski, R., Zaikov, G., Eds.; Nova Science Publishers, Inc.: New York, 2007.
5. Kozlowski, R.; Kozlowska, J; Pawluk, M.; Barriga, J. Potential of lignocellulosic fibrous raw materials, their properties and diversified applications. *Nonlinear Opt., Quantum Opt.* **2004**, *31*, (1–4), 61–89. Special issue Smart and Functional Organic Materials, Proceedings of the NATO Advanced Research Workshop, Bucharest, Romania, June 10–14, 2003.
6. Kozlowski, R.; Manys, S. Coexistence and Competition of Natural Man-made Fibers, Proceedings of the 78th World Conference of the Textile Institute, Thessaloniki Greece, May 23–27, 1997.
7. Kozlowski, R.; Helwig, M. Lignocellulosis Polymer Composites, Proceedings of the 4th International Conference on Frontiers of Polymers Advanced Materials, Cairo, Egypt, 1997.
8. Meadows, D. H.; Meadows, D. L.; Randers, J. *Beyond the Limits: Confronting Global Collapse, Envisioning a Sustainable Future*; Chelsea Green Publishing Company: White River Junction. VT, 1992.
9. Kozlowski, R.; Manys, S. Fiber Plant Should Prolong the Life of Forests, Proceedings of the International Conference on Sustainable Agriculture, Rio de Janeiro, Brazil, March 9–13, 1998.
10. Leão, A. L; Tan, I. H. Tropical Natural Fibers: Potential and Applications in Composites. In *Natural Fibers (Wlokna Naturalne)*, Special ed. 1998/1; Proceedings of the Hemp, Flax and Other Bast Fibrous Plants Production, Technology and Ecology Symposium, September 24–25, 1998; Institute of Natural Fibers: Poznan, Poland, pp 8a–8f.
11. *Renewable Resources: Obtaining, Processing and Applying*; Kozlowski, R., Zaikov, G. E., Pudel, F., Eds.; Nova Science Publishers, Inc.: New York, 2009.
12. Johnson, T. N. F. World fiber demand 1980–2050. *Text. Asia* August **1996**.
13. Kozlowski, R.; Manys, S. Linen Knitted Apparel: Specialty for Hot Climates, Proceedings of the IFKT 39th International Congress, Busto, Italy, October 8, 1998.

14. Kozlowski, R.; Manys, S. Bast Fibers, Proceedings of the 5th Asian Textile Conference, Kyoto, Japan, September 30–October 2, 1999.
15. Kozlowski, R.; Manys, S. Newest Achievements in Curaua Processing and Applications. In *Natural Polymers and Composites*; Mattoso, L. H. C., Leão, A. L, Frollini, E., Eds.; CIP-BRASIL Catalogação-na-publicação, Embrapa Instrumentação Agropecuária: Brazil,2000; Proceedings from the Third International Symposium on Natural Polymers and Composites, ISNaPol/2000 and the Workshop on Progress in Production and Processing of Cellulosic Fibers and Natural Polymers, São Pedro, SP, Brazil, May 14–17, 2000.
16. Kozlowski, R.; Manys, S. The Properties of Liquid Ammonia Treated Linen, Proceedings of the 212th ACSNM, Orlando, FL, August 25–29, 1996.
17. Kozlowski, R; Manys S.; Zimniewska, M.; Mazur E. Linen for Aged/Ill People: New Linen Knitted Antibedsore Bedding. The 1st Nordic Conference on Flax and Hemp Processing, Added Value through Industrial Cooperation, Tampere, Finland, August 10–12, 1998; Institute of Fiber, Textile and Clothing Science, Tampere University of Technology:Tampere, Finland, pp 275–276.
18. Zimniewska, M.; Kozlowski, R. Natural and man-made fibers and their role in creation of physiological state of human body. *Mol. Cryst. Liq. Cryst.* **2004**, *418*, 113–103; Proceedings of the 7th International Conference on Frontiers of Polymers and Advanced Materials, ICFPAM 2003, Part IV of IV, Bucharest, Romania, June 10–14, 2004.
19. Kozlowski, R.; Helwig, M. Critical Look on Cellulose Modification, Proceedings of the International Symposium on Cellulose Modification, Honolulu, HI, December 1996.
20. Kozlowski, R.; Batog, J.; Konczewicz, W.; Mackiewicz-Talarczyk, M.; Muzyczek, M.; Sedelnik, N.; Tanska, B. Enzymes in bast fibrous plant processing. *Biotechnol. Lett.* **2006**, *28* (10), 761–765.
21. K. Bujnowicz; R. Kozlowski; J. Pielichowski; Z. Wirpsza Polymers from Natural Fatty Acids. In *Modern Polymeric Materials for Environmental Applications*; Vol. 1; K. Pielichowski, Ed.; 1st International Seminar, Krakow, Poland, 2004.
22. Kozlowski, R.; Bujnowicz, K. Polymers and Chemicals from the Agro-Resources. In *Renewable Resources: Obtaining, Processing and Applying*; Kozlowski, R., Zaikov, G. E., Pudel, F., Eds.; Nova Science Publishers, Inc.: New York, 2009.

Chapter 4

Immobilized Flavourzyme on Chitosan Beads for Seasoning Sauce Production: Covalent Binding vs Entrapment

Nanthiya Hansupalak,[1,*] Parichart Kitsongsermthon,[1] and Ratana Jiraratananon[2]

[1]Department of Chemical Engineering, Faculty of Engineering, Kasetsart University, Bangkok, 10900 Thailand
[2]Department of Chemical Engineering, King Mongkut's University of Technology Thonburi, Bangkok, 10140 Thailand
[*]fengnyh@ku.ac.th

Flavourzyme was immobilized on chitosan beads using two methods, covalent and entrapment, in which beads were identically prepared. Optimum conditions for covalently binding enzyme on beads activated by glutaraldehyde were examined prior to the study of pH- and thermal- stabilities of immobilized enzymes. The quality of seasoning sauce produced by using immobilized enzymes was measured in terms of the total amino acid nitrogen amount. Both immobilizing techniques improved pH- and thermal-stabilities of free protease. Though the covalent immobilization showed the highest loading efficiency, it yielded the lowest enzymatic activity. Nevertheless due to being the most operational stabilities, covalently bound enzymes gave the highest amount of amino acid nitrogen, which was also greater than that from the similar process where free enzymes (Flavourzyme and amylase) were used, suggesting the potential of using immobilized Flavourzyme for the process.

Introduction

Commercial Flavourzyme is a fungal protease/peptidase complex produced from nongenetically modified *Aspergillus oryzae* and contains both endoprotease and exopeptidase activities, which can digest peptide bonds. The enzyme mixture also contains other substances such as salts and stabilizing agents, but it is nontoxic to human. Traditionally brewed soy sauce is made by mixing soybeans and *Aspergillus oryzae* strains, which then release protease enzymes to breaks down soybean proteins into shorter peptide chains and amino acids, contributing the meat-like, unique flavor of soy sauce. To shorten the processing time from months to days, hydrochloric acid is used to hydrolyze proteins, but an inferior taste and undesired and carcinogenic byproducts, i.e. chloropropanols, occur (*1–3*). Recently it has been found that protease enzymes from crude (or commercial) Flavourzyme can replace hydrochloric acid and yields the product called 'seasoning sauce', containing very small (acceptable) amount of carcinogenic byproducts, but having a similar flavor to the brewed soy sauce (*4*). Pure enzyme is, however, intrinsically unstable and costly due to its production and complicated purification. Immobilizing enzyme onto suitable supports, stable in the medium and large enough to prevent the enzyme loss, is one way to lower the production cost as it facilitates the separation of enzyme from the product and allows the enzyme to be used repeatedly in addition to enhancement of enzyme stability.

Immobilization technique is not new to the soy sauce production. For instance, immobilization of protease, protease-producing fungi, glutaminase, and glutaminase-producing yeast in the soy sauce and seasoning sauce productions havae been investigated (*5–10*). In addition, protease immobilization via covalent, entrapment, or physical adsorption on several supports have been examined widely for various purposes such as production of casein hydrolysates, peptide purification, and protein digestion in proteomics (*11–15*). There are also reports on Flavourzyme immobilization via encapsulation, covalent, and electrostatic adsorption on numerous supports such as sodium alginate/starch mixture, Lewatit R258-K, and glyoxyl-agarose (*16–18*). Nonetheless, none of those works are related to the immobilization of crude Flavourzyme on weak-acid and high-ionic-strength resistant chitosan beads, specifically designed for the seasoning sauce making. In addition this current work prepared chitosan beads for covalent and entrapment in the exact same way, employing ionic interactions between chitosan and sodium tripolyphosphate to solidify chitosan beads (*19*). Therefore, comparison of the immobilized enzyme activities could be made without interferences from the bead preparation. The chitosan is food grade and thus safe for utilizing in the sauce production.

The main objectives of this work was to investigate the possibility of using immobilized Flavourzyme to produce the seasoning sauce. Two immobilization techniques, Covalent and entrapment, were compared. For covalent immobilization, optimum conditions for activating chitosan support using glutaraldehyde and for binding protease on the activated support were examined first so that optimally immobilized Flavourzyme could be employed in the sauce making. Glutaraldehyde is used as a crosslinking molecule to form

Schiff bases with a chitosan mer at one end and with an enzyme molecule at the other end (*20*). Also total amino acid nitrogen was used as a key parameter reflecting the sauce quality.

Experimental

Materials

Flavourzyme protease (500 unit/g or U/g) from Novozymes was a gift from East Asiatic (Thailand) and chitosan (deacetylation degree of 90 and MW~500k) from Elan Corp. (Thailand). All chemicals were used directly without further purification.

Protease Immobilization

Optimum chitosan concentration for entrapping enzymes has been investigated (*21*, *22*) and the 2%w/v chitosan solution is found to be appropriate because the fully formed bead is not too dense to obstruct mass transfer of substrate and products or too loose to hold enzymes. *Covalent Method*: 10 ml of the chitosan solution, made of 2%w/v of chitosan dissolved in 1%v/v acetic acid, was dropped into 0.13 M sodium tripolyphosphate (TPP) solution, which was prepared in 0.1M sodium phosphate buffer (pH7), through a syringe with a 22G needle. After the chitosan beads being cured for 75 min in the same TPP solution, the solution was decanted, and the beads were washed twice using 0.1M sodium phosphate buffer solution (pH7). Next, cleaned beads were agitated in 20 mL of glutaraldehyde solution at pH 7 (prepared in 0.1M sodium phosphate buffer) at room temperature (~28°C) and 220 rpm and later were washed with 0.1M sodium phosphate buffer solution. They were then soaked in 20.5 ml of 0.5M sodium phosphate buffer solution, containing 0.5 ml of Flavourzyme, and shaken at 160 rpm for 3 hr. pH and temperature for the immobilization were varied between 5 – 9 and 30 – 60°C, respectively. Afterwards the beads were stored at 4 °C for 18 hr and then washed in 2M NaCl and deionized water, respectively. *Entrapment*: The same method was applied, except that the chitosan solution was already well mixed with 0.5 ml of Flavourzyme prior to the solidification in the TPP solution. After being immersed in the same TPP solution for 75 min., the beads were washed twice using 0.1M sodium phosphate buffer solution (pH7) and ready to use.

Making Seasoning Sauce

The seasoning sauce was made according to (*4*) with slight modification (the total hours of 17.5 hr was still unchanged). Briefly, 20 ml of defatted soybeans were soaked and shaken in 100 ml distilled water at 220 rpm and 50°C for 4.5 hr. Then optimally immobilized protease was added and the shaking was continued for additional 13 hr at the same temperature. This process was performed in distilled water (pH~7) without buffering. The liquid and solid portions were separated by vacuum filtration for further characterization. Amino acid nitrogen amount in the liquid portion was characterized using a FP-528 LECO instrument (USA).

Protease Assay and Protein Amount

Protease assay: A mixture of 1 ml of 1.5%w/w casein in 0.1M sodium phosphate buffer solution and approximately 0.15 g of beads containing enzyme (or 1 ml of the liquid containing free enzyme) was allowed to react for 10 min at 160 rpm and 40°C. The tyrosine occurred in the solution was quantified by adding 5 ml of Na_2CO_3 and 1 ml of Folin reagent prior to the absorbance measurement at 660 nm using a Shimadzu UV-visible spectrophotometer UV160A (*23*). The activity (U/mg) was expressed as the amount of tyrosine (μmol) produced per reaction time (min) and protein amount. *Protein amount*: 0.6 ml of the liquid part containing enzymes, obtained from experiments, was mixed with 1.5 ml of Coomassie Brilliant Blue G. at room temperature. The UV absorbance of the solution was measured at 595 nm (*24*).

pH- and Temperature- Stabilities

pH-stability was performed by soaking (free or immobilized) enzymes in 0.1M phosphate buffer (ionic strength = 0.05M) at 50°C and various pH for 15 min, prior to the activity measurement. Similarly, the temperature stability was carried out by incubating (free or immobilized) enzymes in 0.1M phosphate buffer (ionic strength = 0.05M) at pH 7 and various temperature for 15 min. Note that optimum activation and immobilization conditions were utilized to covalently immobilize Flavourzyme on chitosan beads.

Results and Discussion

Optimum Conditions for Bead Activation

The covalent bonds between a support and an enzyme can strongly affect enzyme conformation and in turn its activity. Thus it is necessary to find optimum conditions to activate chitosan beads with glutaraldehyde at pH 7 before binding with enzymes. Figure 1 shows that enzymatic activity increases with glutaraldehyde concentration when the concentration is less than 0.1%v/v, after which the activity reduction is observed instead. This is typical for covalent immobilization (*25–27*). The number of covalent bonds formed between the support and enzymes through glutaraldehyde molecules plays an important role here. At low glutaraldehyde concentration, the amount of bonds per one enzyme molecule is too low to hold still an enzyme (single point immobilization), resulting in the leakage, which can be prevented by raising the glutaraldehyde concentration. However, at too high concentration, there are too many bonds per one enzyme (multipoint immobilization), distorted conformation, or enzyme denaturation, occurs. In this work external mass transfer effect may be another cause as high enzyme immobilized amount on the bead surface was observed (results are not shown here).

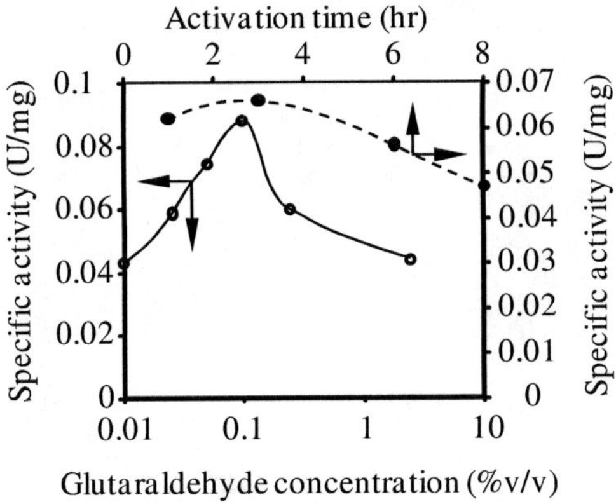

Figure 1. Optimum conditions for activating chitosan beads using glutaraldehyde: activationg time (•; dash lined was obtained by the polynomial regression and R^2 =1; 0.05%v/v glutaraldehyde conc. was used) and glutaraldehyde concentration (○; activation time of 2.5 hr. was used). The immobilization pH and temperature were pH 7 and room temperature, respectively.

At a fixed glutaraldehyde concentration of 0.05%v/v where the number of bonds per one enzyme is small, the low enzymatic activity is observed. Varying the activation time, or exposure time to the glutaraldehyde solution, can alter the enzymatic activity as illustrated in Figure 1. Similarly, the rise in the amount of covalent bonds formed per one enzyme, resulting from the increment of the activation time, may account for the change in the activity.

In Figure 1 the activation conditions yielding the highest Flavourzyme activity are an activation time of 2.5 hr and 0.1%v/v glutaraldehyde concentration solution. Note that the optimum activation time was found by using polynomial regression, yielding R^2 of 1. The highest specific activity at the optimum conditions (0.089 U/mg protein, equivalent to 3.54 U/ml solution) is much smaller than one reported for thiol protease immobilized on glutaraldehyde-activated chitosan beads (32 U/ml) (*27*). This may be explained by the mass/volume ratio of chitosan to the enzyme solution in the current work being 16 times higher, resulting in much lower enzyme loading on chitosan beads. However the specific activity obtained from this current work is still comparable or somewhat higher than what one reported for crude Flavourzyme immobilized on Diaion HP20, glass beads, and silica gel 60 (*17*).

Optimum Conditions for Enzyme Immobilization by Covalent Binding

pH and temperature can influence enzyme conformation, and unsuitable conformation of enzyme can cause the lower enzymatic activity (*28*). It is thus necessary to find optimum pH and temperature for a specific immobilization. It should be noted that for the entrapment the polymeric support should entrap free enzyme as it remains in the active conformation and thus the optimum pH and temperature must be the same as those of free enzymes.

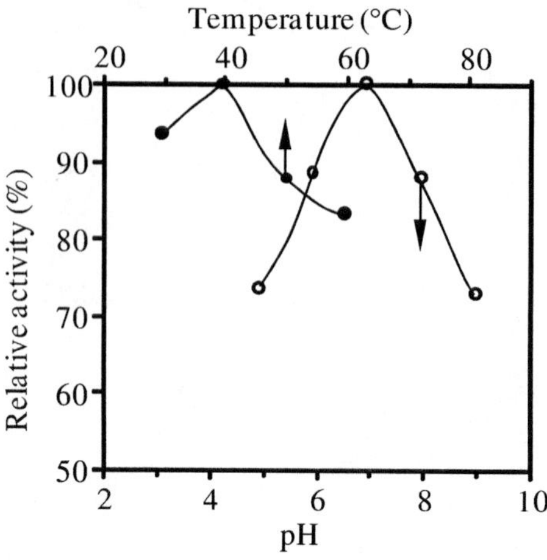

Figure 2. Optimum conditions for Flavourzyme immobilization onto optimally activated chitosan beads: pH (○; constant 30°C) and temperature (●; constant pH 7).

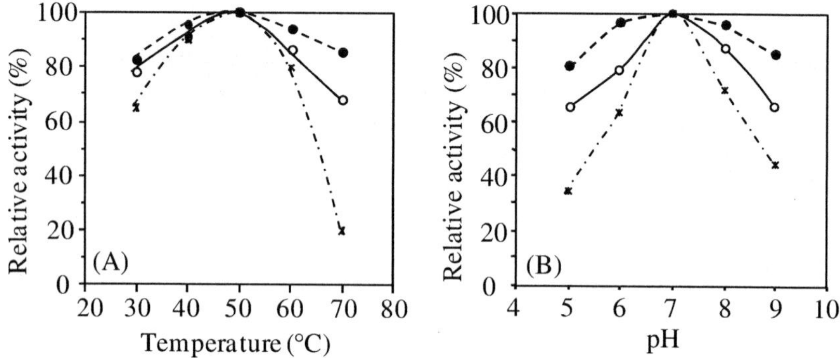

Figure 3. (A) Temperature- and (B) pH-stabilities of free (∗), entrapped (○), and covalently bound (●) enzymes.

For Flavourzyme immobilization on chitosan beads activated using aforementioned optimum conditions, the immobilization conditions yielding the highest activities are pH 7 and 40°C (Figure 2). Note that for this purpose all experiments were conducted in buffer solutions, of which ionic strength was 0.05 M. The optimum pH and temperature found herein correspond to the optimum values suggested by the Novozymes product sheet for neutral free enzymes (*24*), implying the enzyme's conformation unaffected by the present covalent immobilization conditions.

For similar protease immobilization on chitosan beads, Sangeetha and Abraham (*15*) observed the shift in the optimum pH and temperature to higher pH and lower temperature, which implies the conformational change upon the immobilization. This is probably due to his harsher condition, i.e., high glutaraldehyde concentration (2%v/v) such that all amino groups on the enzyme surface are probably linked with glutaraldehyde. Thus untouched acidic groups ionizing at high pH are responsible for shifting the optimum pH to the alkaline region. In contrast when dilute glutaraldehyde concentration solution(0.05 – 1% v/v) is used, the shifting of the optimum pH towards higher pH is not seen (*27*, *29*).

Thermal and pH Stabilities

Enzyme pre-incubation in a phosphate buffer for 15 min at a constant pH or temperature was exercised to observe temperature- and pH-stabilities of enzymes. In Figure 3A, the highest activities obtained from both immobilizations are achieved at the same temperature (50°C) as that of free enzyme, and covalently bound enzyme is the most thermal-stable over a temperature range of 30 – 70°C. In fact both immobilization types improve temperature stability of enzyme, as seen elsewhere (*27*, *29*). The greater temperature stability is due to the crosslinks between the support and enzyme and chitosan matrix that protects immobilized enzymes from heat.

Figure 3B displays both immobilizations expanding pH stability of enzymes to both acidic and basic ranges. The covalently attached enzyme shows the greatest pH stability, reflecting the strongest bonds between the support and enzyme that can maintain the active conformation of enzyme in this pH range. The enhancement of pH stabilities after protease's immobilization on chitosan beads have been observed (*15*, *27*, *30–32*), but optimum temperature and pH values are different depending on the protease types, chitosan concentration and molecular weight, and immobilization techniques.

Loading Efficiency and Quality of the Seasoning Sauce

Enzyme losses during the covalent and entrapment immobilizations are summarized in Table I as mean ± standard deviation. The covalent immobilization retains more enzyme than the entrapment. In the entrapment process, the greatest enzyme loss occurs during the chitosan solidification in TPP solution because the beads are not fully formed, as reported elsewhere (*21*). Free Flavourzyme expresses the highest activity, followed by the entrapped and covalently bound enzymes, respectively. That is because immobilization, especially the covalent, hinders enzyme's ability to combine with substrate (casein). In addition the existence of external and internal mass transfer of substrates, which has already been proved (*32*), leads to a decrease in enzymatic activity, as reported elsewhere (*33*).

The covalently bound enzyme can, however, produce the seasoning sauce which contains the largest concentration of amino acid nitrogen, which represents the quality of the seasoning sauce (*2*, *4*, *10*). It is attributed to the ability of the covalently bound enzyme to withstand the operation conditions better than the entrapped and free enzymes, as discussed previously.

Table I. Enzyme Loading Efficiency and Amino Acid Nitrogen Amount[a]

Immobilization Method	*Initial Protein Amount (mg)*	*Protein Losses (mg)*			*Loading Efficiency (%)*	*Specific activity (U/mg)*	*Total amount of amino acid nitrogen produced (g/dm³)*
		In TPP	*Unbound enzymes*	*In buffer wash*			
Covalent	17.44± 0.94	-	3.64± 0.18	0.49± 0.03	76.3± 0.90	0.086	21.5
Entrapment	17.44± 0.94	9.19± 0.24	-	1.15± 0.03	40.7± 1.40	0.093	15.1
Free enzyme[b]	-	-	-	-	-	0.48	4.1

[a] NOTE: All experiments were repeated thrice. [b] obtained from (*4*).

The amino acid nitrogen amount obtained from employing immobilized enzymes is greater than that obtained from free enzymes (*4*). It should be noted that (i) Sahasakmontri (*4*) utilized both free Flavourzyme and amylase to produce the sauce and the soaking time of 30min before adding enzymes and (ii) defatted soybeans and protease in his work and the current paper came from the same sources. Therefore, the advantage of the current work is that only Flavourzyme is applied (though the longer soaking time before adding protease was needed) and yet can produce higher amino acid nitrogen amount.

Conclusions

Optimized entrapment conditions reported elsewhere were applied herein. For the covalent method, to obtain a high activity value chitosan beads must have been activated using 0.1%v/v glutaraldehyde for 2.5 hr. before immobilization. The optimal immobilization pH and temperature were pH 7 and 40°C, respectively, identical to optimum values of free enzyme, suggesting unchanged conformation during the immobilization. Both covalent and entrapment immobilizations boosted pH and thermal stabilities.

It was found that the covalent method enabled beads to retain enzymes more than the entrapment, because the latter took a few seconds for a chitosan bead to be firm enough to effectively hold enzymes, during which the loss of enzyme occurred. However covalently bound enzyme showed the lowest activity, but due to its most operational stability it could produced the largest amount of amino acid nitrogen, indicative of the better quality seasoning sauce. Surprisingly, the nitrogen amount found in this work, using immobilized Flavourzyme protease alone, was higher than that when using both protease and amylase in the free form, suggesting the potential of applying immobilized Flavourzyme on chitosan in the real sauce production. Nonetheless, to approach the real practice conditions, immobilization of more enzyme amount on chitosan beads should be examined further in order to obtain higher activity.

Acknowledgments

Financial supports from Kasetsart University Research and Development Institute (KURDI), Center of Excellence for Petroleum, Petrochemicals and Advanced Materials, S&T Postgraduate Education and Research Development Office (PERDO), and Thailand Research Fund (TRF, Grant no. MRG4780089) are gratefully acknowledged.

References

1. Hamlet, C. G.; Sadd, P. A.; Crews, C.; Velisek, J.; Baxter, D. E. *Food Addit. Contam.* **2002**, *19*, 619–631.
2. Steinkraus, K. H. *Handbook of Indigenous Fermented Foods (Revised and Expanded)*, 2nd ed.; CRC Press: New York, 1996.
3. *Survey of 3-Monochloropropane-1,2-diol(3-MPCD) in Acid Hydrolysed Vegetable*; Protein Food Surveillance Information Sheet 181; Ministry of Agriculture, Fisheries and Food: London, 1999.
4. Sahasakmontri, K. M.S. Thesis, Kasetsart University, Bangkok, Thailand, 2001.
5. Khare, S. K.; Jha, K.; Gandhi, A. P. *Food Chem.* **1994**, *50*, 121–123.
6. Choi, M.-R.; Sato, N.; Yamagishi, T.; Yamauchi, F. *J. Ferment. Bioeng.* **1991**, *72*, 214–216.
7. Koseko, S.; Hisamatsu, M.; Hirano, K.; Yamada, T. *Nippon Shokuhin Kogyo Gakkaishi* **1994**, *41*, 210–213.
8. Kanematsu, Y.; Kasahara, M.; Hiraguri, Y.; Honkawa, Y. *Nippon Shoyu Kenkyusho Zasshi* **1992**, *18*, 260–269.
9. Motokawa, Y.; Kanematsu, Y.; Kasahara, M.; Yamagata, Y.; Tanaka, T.; Hara, F.; Negi, H. J.P. Patent 64010957, 1989.
10. Motokawa, Y.; Kanematsu, Y.; Yamagata, Y.; Tanaka, T.; Hara, F.; Negi, H. J.P. Patent 62278960, October 20, 1987.
11. Ge, S.-J.; Bai, H.; Yuan, H.-S.; Zhang, L.-X. *J. Biotechnol.* **1996**, *50*, 161–170.
12. Křenková, J.; Klepárník, K.; Foret, F. *J. Chromatogr., A* **2007**, *1159*, 110–118.
13. Nicoli, R.; Rudaz, S.; Stella, C.; Veuthey, J.-L. *J. Chromatogr., A* **2009**, *1216*, 2695–2699.
14. Megias, C.; Pedroche, J.; Yust, M. M.; Giron-Calle, J.; Alaiz, M.; Millan, F.; Vioque, J. *J. Agric. Food. Chem.* **2007**, *55*, 3949–3954.
15. Sangeetha, K.; Abraham, T. E. *J. Appl. Polym. Sci.* **2008**, *107*, 2899–2908.
16. Kailasapathy, K.; Perera, C.; Phillips, M. *Int. J. Food Eng.* **2006**, *2*.
17. Chae, H. J.; In, M.-J.; Kim, E. Y. *Appl. Biochem. Biotechnol.* **1998**, *73*, 195–204.
18. Yust, M. d. M.; Pedroche, J.; Alaiz, M.; Giron-Calle, J.; Vioque, J.; Mateo, C.; Guisan, J. M.; Millan, F.; Fernandez-Lafuente, R. *J. Agric. Food. Chem.* **2007**, *55*, 6503–6508.
19. Hsieh, F.-M.; Huang, C.; Lin, T.-F.; Chen, Y.-M.; Lin, J.-C. *Process Biochem.* **2008**, *43*, 83–92.

20. L′opez-Gallego, F.; Betancor, L.; Hidalgo, A.; Alonso, N.; Fernandez-Lorente, G.; Guisan, J. M.; Fernandez-Lafuente, R. *Enzyme Microb. Technol.* **2005**, *37*, 750–756.
21. Betigeri, S. S.; Neau, S. H. *Biomaterials* **2002**, *23*, 3627–3636.
22. Adriano, W. S.; Mendonc, D. B.; Rodrigues, D. S.; Mammarella, E. J.; Giordano, R. L. C. *Biomacromolecules* **2008**, *9*, 2170–2179.
23. Han, X. Q.; Shahid, F. *Food Chem.* **1995**, *52*, 71–76.
24. Bradford, M. M. *Anal. Biochem.* **1976**, *72*, 248–254.
25. Spagna, G.; Barbagallo, R. N.; Casarini, D.; Pifferi, P. G. *Enzyme Microb. Technol.* **2001**, *28*, 427–438.
26. Jiang, D.-S.; Long, S.-Y.; Huang, J.; Xiao, H.-Y.; Zhou, J.-Y. *Biochem. Eng. J.* **2005**, *25*, 15–23.
27. Bhandari, S.; Gupta, V. K.; Singh, H. *Biocatal. Biotransform.* **2009**, *27*, 71–77.
28. Chaplin, M. F.; Bucke, C. *Enzyme Technology*; Cambridge University Press: Cambridge, U.K., 1990.
29. Altun, G. D.; Cetinus, S. A. *Food Chem.* **2007**, *100*, 964–971.
30. Ha, B.-J.; Lee, O.-S.; Lee, Y.-S. *J. Korean Ind. Eng. Chem.* **1996**, *7*, 186–193.
31. Hayashi, T.; Ikada, Y. *J. Appl. Polym. Sci.* **1991**, *42*, 85–92.
32. Yaosong, N.; Hansupalak, N. *Fungal Protease Covalently Bound on Chitosan Beads: Microscopy and Mass Transfer Aspects*. 2nd Polymer Graduate Conference of Thailand, May 21–22, 2009; Chulalongkorn University: Thailand, 2009; in preparation.
33. Benyahia, F.; Polomarkaki, R. *Process Biochem.* **2005**, *40*, 1251–1262.

Polymers in the Nano/Bio/Info Technology Revolution

Chapter 5

Triphilic Block Copolymers: Synthesis, Aggregation Behavior and Interactions with Phospholipid Membranes

S. O. Kyeremateng, C. Schwieger, A. Blume, and J. Kressler*

Department of Chemistry, Martin Luther University Halle-Wittenberg, D-06099 Halle (Saale), Germany
***Corresponding author. E-mail: joerg.kressler@chemie.uni-halle.de.**

Novel PPO-based amphiphilic diblock copolymer and triphilic multiblock copolymer analogues of the architectures BA, CBA, and CABAC have been successfully synthesized by combination of atom transfer radical polymerization (ATRP) and copper(I)-catalyzed alkyne-azide cycloaddition (CuAAC) 'click' post-polymerization reaction. The A, B, and C components of the block copolymers are formed by hydrophilic poly(glycerol monomethacrylate) (PGMA), lipophilic poly(propylene oxide) (PPO), and perfluoroalkyl segment, respectively. Their critical micelle concentrations in water are determined from surface tension measurements. Their aggregation behavior in aqueous medium is studied by temperature-dependent ^{1}H- and ^{19}F NMR spectroscopy. The presence of the perfluoro segments within the micelle cores of the triphilic block copolymers enhanced 3-4 folds their solubilization capacity of tetrafluorobenzene compared to micelle cores without the perfluoro segments. It is found that the triphilic copolymer inserts into lipid membranes and induces lateral demixing between the two homologue lipids in the temperature range of the phase transition of the lipids.

Introduction

Properties of polymers intended for advanced applications have continuously been expanded through block copolymerization. Consequently, novel block copolymers are being created and characterized extensively in the field of polymer synthesis. The most widely studied and industrially applied amphiphilic block copolymer of the type ABA is poly(ethylene oxide)-*b*-poly(propylene oxide)-*b*-poly(ethylene oxide), PEO-PPO-PEO, commonly referred to by the generic name Poloxamer (*1*, *2*). For most parts of the behavior of the block copolymer in water, PPO and PEO behave as the thermoresponsive-lipophilic and hydrophilic blocks, respectively. They find widespread application, partially because of their commercial availability, in investigations dealing with drug delivery of poorly water-soluble drugs inside micelles (*3*, *4*) or hydrogels (*5*) and sensitization of multi-drug resistance (MDR) cancer cells (*6*). The chemo-sensitizing role played by Poloxamers in MDR cancer treatments is attributed to their ability to increased microviscosity ('fluidized') of the membranes of cancer cells, while, in contrast, decreased the microviscosity ('solidified') of membranes of the normal blood cells (*6*, *7*).

Despite all these advantages, it is worth mentioning that Poloxamers still present a number of drawbacks that depending on the specific case could limit the application. The most important has to do with the generally low stability of drug-carrying Poloxamers micelles upon dilution in the bloodstream. This phenomenon is more crucial in derivatives with increased block length ratio of PEO/PPO (*4*). The ability of the micelles to remain stable in the bloodstream upon dilution very much depends on the critical micellization concentration (cmc) of the block copolymer. Micelles formed from block copolymers with low cmc are more stable upon dilution and vice-versa (*4*, *6*). Also, Poloxamers lack the ability to be retained up to surface pressures relevant for some biological membranes such as the blood-brain barrier and erythrocyte membrane (*8*).

In this context, it will be desirable to create analogous block copolymers with properties that can overcome some of the drawbacks associated with Poloxamers. It is along this direction that motivated this research to substitute the PEO block with poly(glycerol methacrylate), PGMA as the hydrophilic block, thus, creating amphiphilic PPO-based block copolymers analogous to Poloxamers and PPO-PEO diblock copolymer. The PGMA block was chosen as a suitable substitute because of its biocompatibility, and has already found commercial application in contact lens (*9–11*). Moreover, recent reports have also pointed out that replacement of PEO with hydroxyl-bearing hydrophilic polymers leads to comparatively lower cmcs (*12*, *13*). Atom transfer radical polymerization (ATRP) using alkyl bromide derivative of poly(propylene oxide) as macroinitiator was employed as the main synthetic technique. Furthermore, we also introduce short perfluoroalkyl segments at the ends of the copolymer chains through copper(I) azide-alkyne cycloaddition (CuAAC) 'click' reaction (*14*, *15*). Thus, a hydrophilic A, a lipophilic B, and a fluorophilic C component (which is neither hydrophilic nor lipophilic) offer a strong three-fold philicity, i.e., triphilic system. The introduction of the perfluoro segments is expected to further lower the cmc of these block copolymers as well as create the potential use of their micelles as delivery vehicles for lipophilic and/or fluorophilic agents (*16*). Particularly, we demonstrate this potential via the enhanced uptake of a perfluorocarbon-based hydrophobe by these micelles. Similar to halogenated drugs, it is also expected that the presence of the perfluoro segment, in addition to the lipophilic block, can enhance membrane binding and permeation as schematically depicted in Figure 1 (*17*).

In this chapter we describe the synthesis, aggregation behavior, and interaction of these novel block copolymers with phospholipid membranes.

Experimental Section

Materials

All chemicals were bought from Sigma-Aldrich unless otherwise stated. Dihydroxy-terminated poly(propylene oxide) (HO-PPO-OH) [M_n (SEC) ~2000 g mol^{-1}, M_n (^{1}H NMR) ~1600 g mol^{-1}] was used as received. The monomer, solketal methacrylate (SMA), used for the polymerizations was synthesized by a procedure reported elsewhere (*18*). The alkyne terminated perfluoroalkyl segment ($F_9C{\equiv}H$) used in the CuAAC 'click' reaction was also by a procedure reported elsewhere (*14*). Triethylamine (Et_3N) (99.8%), dimethylformamide (DMF) (99.8%) and anisole (Alf-Aesar, 99%) were dried by standard procedures. Ethanol (99.8%), *n*-hexane (97%), 2-bromoisobutyryl bromide (BIB) (98%) diethyl ether (98%), and 1,4-dioxane (99%) were used as received. Tetrahydrofuran (THF) (99.5%) was distilled from potassium hydroxide and stored over molecular sieve. Copper bromide (CuBr) (99%) was purified by stirring in glacial acetic acid under nitrogen for 24 h to dissolve the Cu(II) species, filtered, washed several times with ethanol and dried under vacuum. N-ethyl-diisopropylamine (DIPEA) (98%), tris-(benzyltriazolylmethyl)amine (TBTA) (97%), sodium azide (NaN_3) (99.5%), and 2,2′-bipyridine (Bi-py) (Merck, 99.5%) were used without further purification. 1,2-Dipalmitoyl-*sn*-glycero-3-phosphocholine (DPPC) and 1,2-dimyristoyl-*sn*-glycero-3-phosphocholine (DMPC) were purchased from Genzyme Pharmaceuticals (Liestal, Switzerland). DPPC-d_{62}, was purchased from Avanti Polar Lipids (Alabaster, Alabama, USA). The lipids were used without further purification.

Instrumentation

Size exclusion chromatography (SEC) measurements were performed in tetrahydrofuran (THF) at room temperature on Viscotek VE 2001 column equipped with RI detector Viscotek 3580. Polystyrene was used as the calibration standard. ^{1}H and ^{19}F NMR spectra were recorded from 25 to 60 °C in D_2O and at 25 °C in DMSO-d_6 using Varian magnetic resonance equipment with Gemini 2000 spectrometers operating at 200 and 400 MHz for measurements in D_2O and DMSO-d_6, respectively. Polymer solutions of concentration 7 g/L were employed. The solutions were prepared by dissolution of appropriate amounts of polymer in solvent followed with slight agitation. Surface tension (γ) of the aqueous solutions of the samples at different polymer concentrations was measured by the Wilhelmy plate method using the automated DCAT11 tensiometer (Data Physics

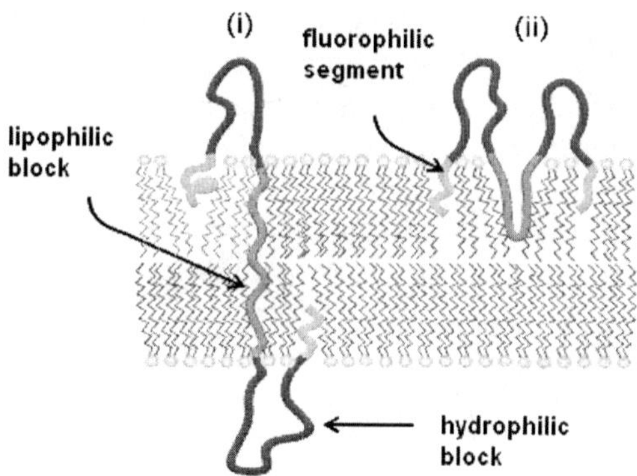

Figure 1. Interaction modes of a triphilic block copolymer with a model lipid membrane: (i) trans-membrane spanning; (ii) partial insertion.

Instruments GmbH, Filderstadt, Germany). Differential Scanning Calorimetry (DSC) was performed with a Microcal VP-DSC (MicroCal Inc., Northhampton, USA). In all experiments we used a heating rate of 1 K/min and a time resolution of 4 s. Lipid vesicles and polymer samples were prepared separately and mixed directly before the measurement. The lipid concentration was always 1 mmol/L. The polymer was prepared in a concentration to give the desired polymer content after adding it in a 1/1 (vol/vol) mixing ratio to the lipid vesicle suspension. The mixing order was always polymer to lipid. Reference was degassed milliQ water. Repeated up- and down scans were performed for each sample until the heat profiles were reproducible. All presented curves are baseline corrected.

Preparation of Lipid Vesicles

Lipids were dispersed in aqueous solution by heating over the phase transition temperature and repeated vortexing. Large unilamelar vesicles (LUV) were prepared by repeated extrusion (15 times) through a 100 nm polycarbonate membrane using a Liposofast-Extruder (Avestin) at 10 °C above the gel to liquid-crystalline phase transition temperature to give vesicles 100 nm in diameter. The size was controlled by DLS messurements. For the preparation of mixed lipid vesicles the DMPC and DPPC-d_{62} were dissolved in $CHCl_3$. The $CHCl_3$ solutions were mixed in equimolar ratio before the solvent was evaporated by gentle heating in a N_2-stream. The obtained mixed lipid films were then dispersed in H_2O and further treated as described above. The DPPC with perdeuterated acyl chains was used in the mixed vesicles preparation in order to distinguish between DMPC and DPPC in subsequent IR measurments (results not discussed here). The perdeutetration has no influence on the miscibility of the lipids.

Tetrafluorobenzene (TFB) Solubilization Experiments

For solubilization experiments, 5 mg/mL D_2O solutions of the polymers were prepared by stirring for 24 h. The polymer solutions were then transferred into clean NMR tubes and 30 μL of TFB was added followed by strong agitation of the mixture for 1 h. Then the tubes were kept in the dark for at least 3 weeks to allow the excess TFB to phase separate to the bottom of the tube. Carefully, 0.8 ml of the clear top aqueous solutions were removed and analyzed by ^{1}H NMR spectroscopy. The amounts of solubilized TFB were then calculated from the integral of the characteristic 2H signals of the benzyl ring of TFB at 7.01 ppm in comparison to the signal of the backbone methylene group of the hydrophilic PGMA block at 1.86 ppm.

Synthesis of ATRP Macroinitators

α,ω-2-Bromoisobutanoate Poly(propylene oxide (Br-PPO-Br)

In a 100 mL two-necked round bottom flask containing 5 g of HO-PPO-OH (2.5 mmol), 30 mL of toluene was added. Azeotropic distillation of the solution was carried out using the Dean and Stark apparatus. The solution was allowed to cool and 0.76 g (7.5 mmol) of Et_3N added. The flask was then placed in a water-ice bath and 1.72 g (7.5 mmol) of BIB added dropwise and slowly over a 30 min period. Reaction was carried out at room temperature for 48 h. Purification was achieved by first filtering off the $(Et_3NH)^+Br^-$ salt formed, followed by removal of toluene by rotary evaporation under reduced pressure. The polymer was dissolved in 120 mL methanol/water solution and NaOH solution (0.25 N) was added with vigorous shaking until neutral pH. Methanol was removed and the polymer extracted with dichloromethane. The extract was decolorised with activated charcoal, dried with anhydrous $MgSO_4$ and dichloromethane removed under reduced pressure to afford Br-PPO-Br (*14*).

α-Azido-ω-2-bromoisobutanoate Poly(propylene oxide) (N_3-PPO-Br)

Similar to the preparation of the macroinitiator above, N_3-PPO-Br was prepared by first partial acylation of HO-PPO-OH in a molar ratio of [PPO]:[BIB]:[Et_3N]; 1:1.3:1.3. Thin layer chromatography (TLC) showed the product did not contain unacylated PPO. Based on this knowledge, it was estimated from analysis of the ^{1}H NMR spectrum that the product contains about 15 mol% of completely acylated PPO; i.e., 85 mol-% of the PPO is end capped with a free OH group. The terminal Br group of the chains was then replaced with N_3 through azidation reaction with NaN_3 according to a method reported elsewhere (*20*). The terminal free OH of the PPO chains was futher acylated completely with BIB using the same method as mentioned above, but replacing Et_3N with pyridine and dichloromethane with diethyl ether, to afford 85 mol% N_3-PPO-Br, and 15 mol% α,ω-diazido-terminated poly(propylene oxide) (N_3-PPO-N_3) (*14*).

Synthesis of Block Copolymers

The N_3-PPO_{27}-$PSMA_y$ block copolymers were prepared by ATRP of SMA using α-azido-ω-2-bromoisobutanoate poly(propylene oxide) (N_3-PPO-Br) as macroinitiator. The subscript y represents the degree of polymerization as determined by ^{1}H NMR using the relation I_e/I_c x n, with I_e/I_c being the integral ratio of the PSMA backbone and PPO methyl protons peaks, and n equals the degree of polymerization of the PPO block which is 27. The reaction scheme is presented in Figure 2. As a typical example for the general experimental procedure, 16 mg (0.11 mmol) of CuBr and 53 mg (0.34 mmol) of Bi-py were placed in a dry schlenk flask equipped with a stir bar. The flask was evacuated under high vacuum and back-filled with nitrogen three times before leaving it under nitrogen. 1 mL of previously degassed anisole was introduced into the flask via a nitrogen-purged syringe. The solution was stirred at room temperature for 30 min to enable homogenization and formation of the catalyst-ligand complex. This was followed by addition 525 mg (2.6 mmol) of degassed SMA via a nitrogen-purged syringe. 300 mg (0.11 mmol) of the heterofunctional macroinitiator was dissolved in 1 mL of degassed anisole and introduced into the flask via a nitrogen purged-syringe. Degassing was carried out for 15 min after which polymerization was carried out at 40 °C. After 20 h the flask was opened to air, allowed to cool and excess THF added. The polymer was purified by column chromatography followed by precipitation into *n*-hexane. Note the PPO-based macroinitiators are soluble in *n*-hexane; therefore, the inactive N_3-PPO-N_3 was isolated from the diblock copolymers during the precipitation process. Two block copolymers were prepared with this macroinitiator.

Using α,ω-2-bromoisobutanoate poly(propylene oxide) (Br-PPO-Br) as macroinitiator ATRP of SMA was carried out to yield $PSMA_y$-PPO_{27}-$PSMA_y$ block copolymers as schematically illustrated in Figure 3. The same experimental procedure mentioned above was applied, except, the $[initiator]_0$:$[CuBr]_0$:$[Bi\text{-}py]_0$ ratio used in this case was 1:1:2. The ratio of the monomer to initiator was varied depending on the desired degree of polymerization. Likewise, two polymerization reactions were carried out for 90 min each. The relatively short polymerization time was employed to maintain a high degree of bromine chain-end functionality (*19*). After purifying the polymers, substitution reaction of the bromine chain-end functionality with azido functionality was performed NaN_3 with in DMF for the 24 h, to give N_3-$PSMA_y$-PPO_{27}-$PSMA_y$-N_3 (*20*)

Through CuAAC 'click' reaction of the azido end-groups with $F_9C{\equiv}H$, one of the N_3-PPO_{27}-$PSMA_y$ copolymers and the two N_3-$PSMA_y$-PPO_{27}-$PSMA_y$-N_3 copolymers were end-capped with perfluoroalkyl segments, yielding, F_9-PPO_{27}-$PSMA_y$ and F_9-$PSMA_y$-PPO_{27}-$PSMA_y$-F_9 respectively. The general experimental procedure for one of the reactions is described in the following. In a schlenk flask containing 570 mg (0.03 mmol, M_n ~18700 g mol^{-1}) of N_3-$PSMA_y$-PPO-$PSMA_y$-N_3 and 72 mg of $F_9C{\equiv}H$ (0.12 mmol), 5 mL of freshly distilled THF was added and stirred for complete dissolution. The solution was degassed for 15 min with nitrogen followed by addition of 3 mg CuBr (0.02 mmol), 16 mg DIPEA (0.12 mmol) and 3.2 mg TBTA (0.006 mmol). Further degassing was carried out for additional 10 min and the flask placed in an oil bath at 50 °C for 20 h (*21*). Completion of the reaction was confirmed by FT-IR spectrum which showed complete disappearance of the azide band at 2112 cm^{-1}. In addition ^{19}F NMR results also confirmed signals from the fluorine moieties. Purification of

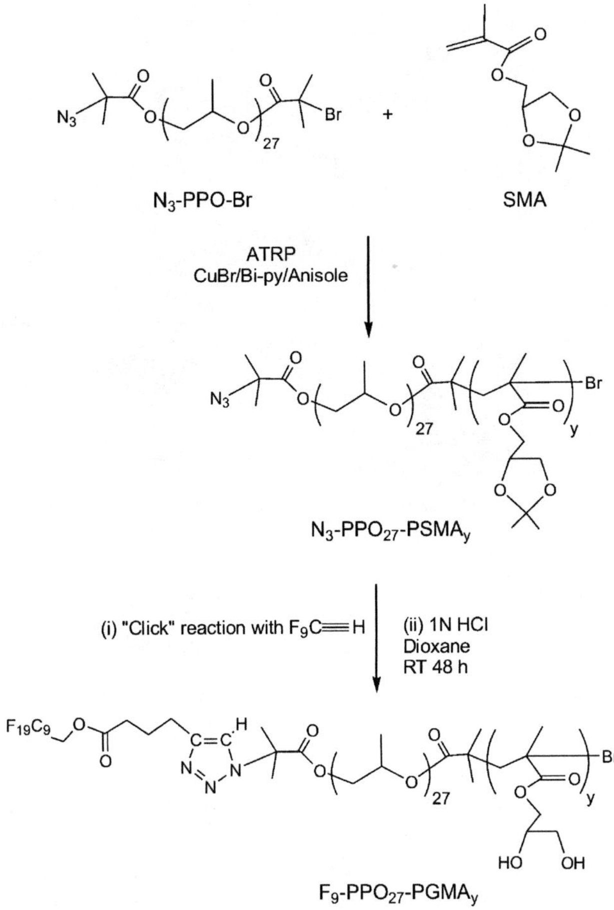

N_3-PPO-Br + SMA

ATRP
CuBr/Bi-py/Anisole

N_3-PPO$_{27}$-PSMA$_y$

(i) "Click" reaction with $F_9C{\equiv}H$ (ii) 1N HCl Dioxane RT 48 h

F_9-PPO$_{27}$-PGMA$_y$

Figure 2. Synthetic route to triphilic F_9-PPO$_{27}$-PGMA$_y$ block copolymers.

the product was by column chromatography follow by precipitation into excess *n*-hexane.

The ketal functions on the SMA units of all synthesized block copolymers were then completely hydrolyzed to give water-soluble glycerol monomethacrylate (GMA) units as indicated in the last step of Figures 2 and 3. The ^{1}H-NMR spectra showed the complete disappearance of the CH_3 (1.17 -139 ppm) corresponding to the two pendant methyl groups on the SMA unit upon complete hydrolysis. The OH functional groups formed appeared at 4.69 and 4.95 ppm in the spectra (*18*). Thus, one amphiphilic and three triphilic block copolymers were finally obtained. Table 1 list the number average molar masses (M_n) and polydispersity indices (M_w/M_n) obtained from ^{1}H NMR and

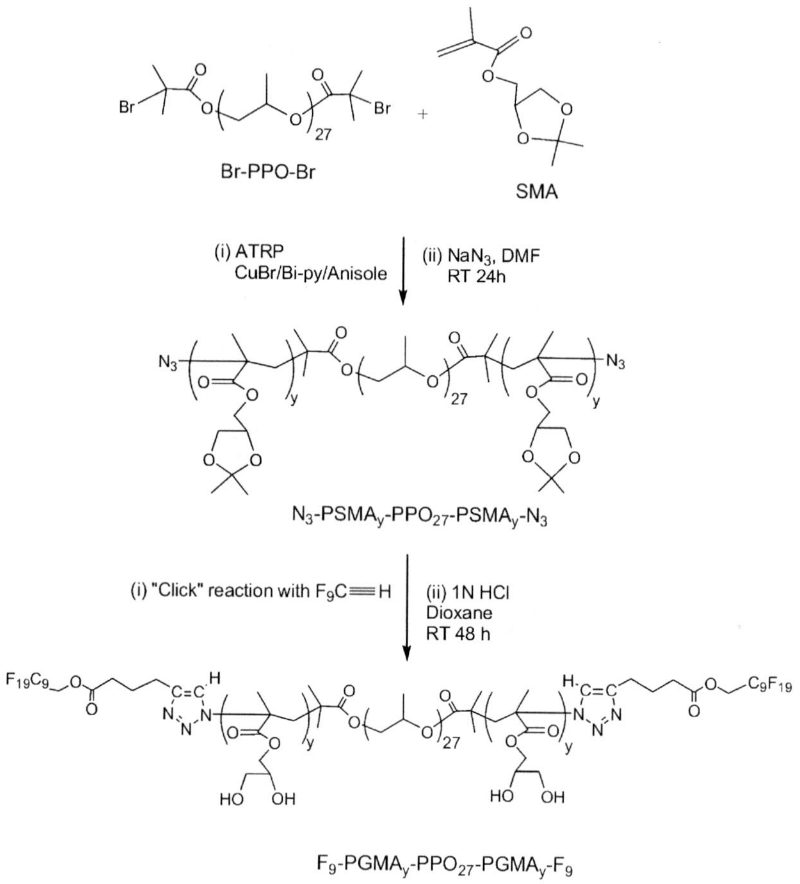

Br-PPO-Br + SMA

(i) ATRP CuBr/Bi-py/Anisole (ii) NaN_3, DMF RT 24h

N_3-PSMA$_y$-PPO$_{27}$-PSMA$_y$-N_3

(i) "Click" reaction with $F_9C{\equiv}H$ (ii) 1N HCl Dioxane RT 48 h

F_9-PGMA$_y$-PPO$_{27}$-PGMA$_y$-F_9

Figure 3. Synthetic route to triphilic F_9-PGMA$_y$-PPO$_{27}$-PGMA$_y$-F_9 block copolymers.

SEC, respectively, for the synthesized block copolymers. Detailed and extensive characterization of the block copolymers is discussed elsewhere (*14*).

Results and Discussion

The aggregation behavior of the block copolymers in aqueous solution is investigated by surface tension measurements and temperature-dependent ^{19}F- and ^{1}H NMR spectroscopy measurements in D_2O. Interactions with phospholipids membranes are also investigated using DSC for one of the triphilic block copolymers. For the discussion part the block copolymers will be referred by their analogous block architectures as given in Table 1.

Table 1. Molar Mass and Polydispersity Indices of Synthesized Block Copolymers

Block copolymer	*Analogous block architecture*	*^{1}H NMR M_n (g mol^{-1})*	*SEC^a M_w/M_n*
N_3-PPO_{27}-$PGMA_{44}$	BA	8800	1.23
F_9-PPO_{27}-$PGMA_{94}$	CBA	17400	1.36
F_9-$PGMA_{24}$-PPO_{27}-$PGMA_{24}$-F_9	CABAC	10500	1.20
F_9-$PGMA_{42}$-PPO_{27}-$PGMA_{42}$-F_9	CA'BA'C	16500	1.50

[a] Obtained from measurements of unhydrolyzed polymers in THF with poly(styrene) calibration standards.

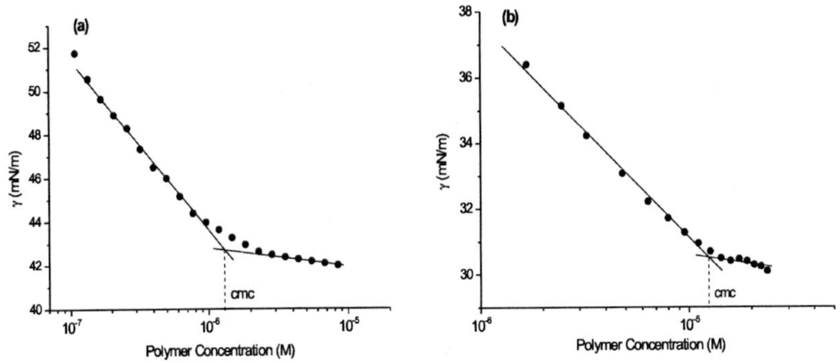

Figure 4. Critical micellization concentration (cmc) determination of (a) CBA and (b) BA from surface tension measurements as a function of concentration at 25 °C.

cmc Determination

For low-molar-mass surfactants or amphiphilic block copolymers that self-assembly in solution, the critical micellization concentration (cmc) is an important physical parameter that characterizes such systems. Surface tension measurement over a wide range of concentration is one of the several methods used for the cmc determination. Therefore, surface tension measurements were carried out on aqueous solutions of the block copolymers in order to obtain information on micelle formation. The surface tensions, γ, were measured as a function of polymer concentrations at 25 °C for BA, CBA, CABAC, and CA'BA'C. Plotting γ versus polymer concentration yields the cmc, indicated by intersection of the extrapolation of the two linear regimes where the curve shows an abrupt change in slope as depicted in Figure 4 for CBA and BA copolymers. The value obtained by this method for CBA and BA are 1.3 and 12.4 μM, respectively. These values are about two orders of magnitude less than those obtained for micellization of PPO-PEO and PEO-PPO-PEO block copolymers of comparable composition at similar temperature (*22, 23*).

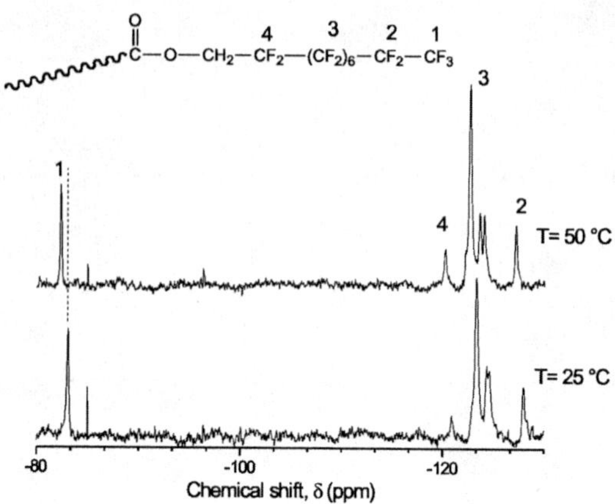

Figure 5. ^{19}F NMR spectra of 7 g/L CABAC in D_2O obtained at 200 MHz for 25 and 50 °C.

Although CBA has a relatively larger hydrophilic PGMA block content, its cmc value is significantly lower than that of BA. This can be attributed to the presence of the highly hydrophobic perfluoro segment at the end of the PPO block in the case of CBA. By the same approach, the cmc values obtained at 25 °C for CABAC and CA′BA′C are 9.5 and 2.5 μM, respectively. Interestingly, CABAC which is more hydrophobic than CA′BA′C yielded a cmc value closer to the BA copolymer even though it contains perfluoro segments.

^{19}F NMR

Due to the linking sequence of the hydrophobic components (C and B), it is obvious that for the CBA architecture both the fluorophilic and lipophilic components should form the core structure when aggregation occurs because they are covalently linked to each other. Considering the particular sequence of the CABAC and CA′BA′C architectures, the hydrophilic blocks are expected to loop to shield the hydrophobic components (fluorophilic and lipophilic) from the aqueous environment when micelles are formed. To ascertain whether the F_9 segments are located within the core or corona of the micelle structures formed, ^{19}F NMR measurements are carried out at 25 and 50 °C on 7 g/L aqueous solutions of the block copolymers. In Figure 5 the ^{19}F NMR spectra of CABAC solution with the corresponding resonance signals assignment is shown.

At 25 °C high-resolution ^{19}F resonance signals can be observed, an indication that the mobility of the fluorine moieties is not restricted. Thus, the F_9 segments do not contribute to the formation of the cores of the micelles. Increasing the temperature to 50 °C records a better signal resolution. Especially, the signal from the CF_2 unit (labeled 4 in Figure 5) closest to the hydrophilic block becomes very prominent. Additionally, the temperature increase causes an increase in the chemical shift of the resonance signals by ~0.7 ppm. In the literature, such increase in chemical shift and signal resolution has been attributed to increasing mobility of the fluorocarbon moieties (*24*, *25*). Thus, the increase in mobility can be attributted to breaking of H-bonding between the coronal PGMA chains which consequently leads to an increase in mobility of the F_9 segments since they are located at the ends of the PGMA blocks (*26*).

CA'BA'C and CBA on the contrary did not register any ^{19}F signal when measurements were carried out under identical experimental conditions. Thus, suggesting that the perfluoro F_9 segments in this case are part of the micelle core. The lack of the F_9 segments within the core of CABAC micelles is due to the inability of their short rigid PGMA blocks to loop (*26*) and may explain the relatively high cmc value of CABAC compared to the other triphilic block copolymers, i.e., CA'BA'C and CBA.

^{1}H NMR

NMR spectroscopy is a powerful tool for investigating mobility of polymer chains in solution. Figure 6 shows the ^{1}H NMR spectra of the methyl protons of the PPO (PPO−CH_3) and the PGMA (PGMA−CH_3) blocks of BA and CBA at 25 °C in DMSO-d_6, a nonselective solvent for both blocks as well as the fluorophilic segment. The peak at 0.76 ppm corresponds to the high content syndiotactic rr triads PGMA methyl protons (rr-PGMA−CH_3), while the peak at 0.93 ppm corresponds to the heterotactic rm triads (rm-PGMA−CH_3) of the same protons (*27*).

Analysis of the syndiotactic sequence content gives 66%, which is similar to values obtained for poly(methyl methacrylate) (PMMA) prepared by radical polymerization (*28*, *29*). The usual splitting of the PPO methyl protons peak at 1.02 ppm is due to *J* coupling of the methyl group to the methine group and can be noticed clearly in both spectra (*30*).

Figure 7 compares the systematic changes in the ^{1}H NMR spectra with temperature for the PPO−CH_3 and PGMA−CH_3 resonance signals of the two copolymers above cmc in D_2O. The temperature-dependent residual HDO resonance signal was corrected for each temperature in accordance with earlier studies by Gottlieb et al. (*31*).

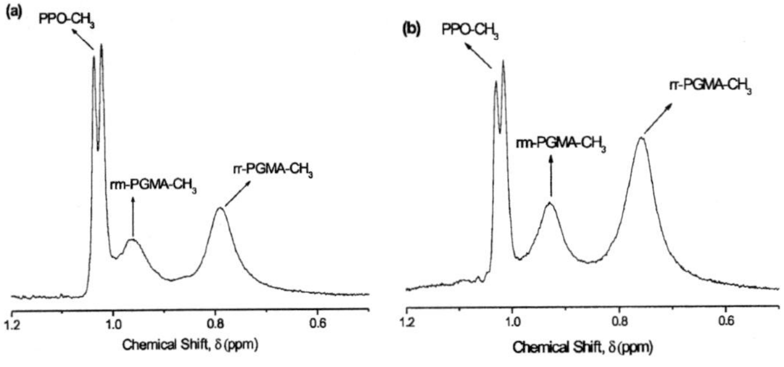

Figure 6. ^{1}H NMR spectra at 400 MHz of 7 g/L of polymer in DMSO-d_6 at 25 °C (a) BA and (b) CBA.

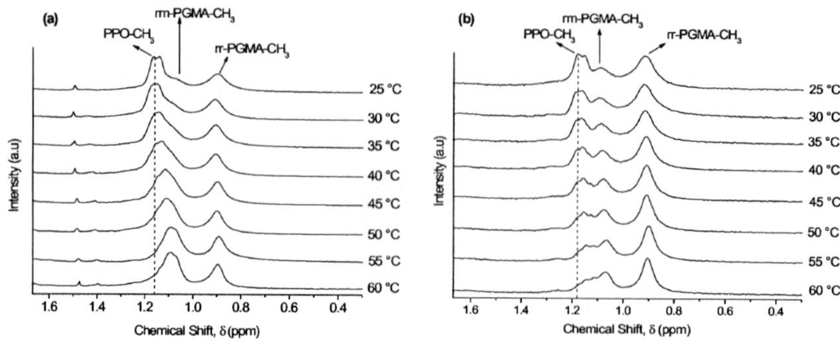

Figure 7. ^{1}H NMR spectra at 200 MHz of 7 g/L of: (a) BA and (b) CBA in D_2O at different temperatures, showing the PGMA-CH_3 and PPO-CH_3 signals. The dashed line represents the position of the PPO methyl protons at 25 °C.

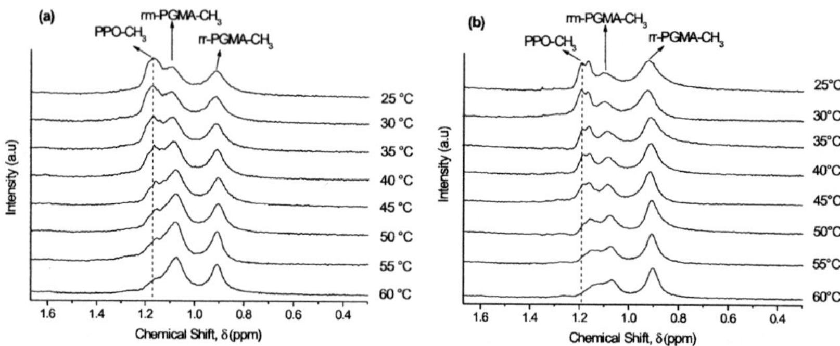

Figure 8. ^{1}H NMR spectra at 200 MHz of 7 g/L of: (a) CABAC and (b) CA'BA'C in D_2O at different temperatures, showing the PGMA-CH_3 and PPO-CH_3 signals. The dashed line represents the position of the PPO methyl protons at 25 °C.

In Figure 7a it is clearly evident that the PPO−CH_3 resonance signal intensity of BA at 25 °C is highly attenuated relative to the PGMA−CH_3 signals when this spectrum is compared to the counterpart spectrum obtained in DMSO-d_6 at the same temperature (see Figure 6a). Also, the usual splitting of the PPO−CH_3 resonance signal has completely disappeared and the signal broadened as well. Attenuation, broadening and disappearance of the splitting of the PPO−CH_3 signal are due to the reduced mobility of the PPO block in D_2O (*30*, *32*). These significant changes in the PPO−CH_3 resonance signal of BA indicate a change in the chemical environment of the methyl protons and signify that the majority of the PPO block is already in the relatively hydrophobic microenvironment of a micelle core (*30*, *32*). Similarly, the PPO−CH_3 resonance signal of the CBA spectrum in D_2O at 25 °C shown in Figure 7b also exhibits features attributable to PPO blocks in the microenvironment of a micelle core.

A clear observation in the ^{1}H NMR spectra of BA in Figure 7a is the shifting of the resonance signal of the PPO-CH_3 upfield with increasing temperature while the rr-PGMA-CH_3 resonance signal remains at approximately the same position.

Eventually, the PPO-CH_3 signal completely overlaps with the rm-PGMA-CH_3 signal, resulting in a broader intense single peak at 1.07 ppm at 60 °C. The upfield shift is due to the change in magnetic susceptibility around the PPO–CH_3 protons owing to the de-shielding effect caused by removal of water molecules around the protons (*32*). Thus, with increasing temperature the PPO core becomes increasingly dehydrated and hydrophobic (*33, 34*). The fact that the PPO-CH_3 signal is still very pronounced at 60°C although it is in the micelle core signifies that the core contains significant amount of trapped-water which is not easily removed. It can be clearly seen from Figure 7b that this overlap effect is less pronounced in the ^{1}H NMR spectrum of CBA at 60 °C. This is because the presence of the highly hydrophobic F_9 segments within the micelle cores of CBA creates a well-dehydrated solidlike environment within the core. Hence, the signal response of the PPO–CH_3 protons in this case is relatively weak. Likewise, at 60 °C there is an overlap of the PPO–CH_3 and the rm-PGMA–CH_3 signals, but the cumulative effect is very weak compared to CBA. In fact, this simple analyses of the temperature-dependent ^{1}H NMR spectra of BA and CBA conveniently upholds the existence of significant amount of water within micelle cores composed of only PPO as has been proven by SANS measurements (*35–38*). Similar ^{1}H NMR spectra observations regarding partially hydrated PPO cores of PPO-PGMA micelles in aqueous solution at high temperatures have been reported by Save et al. (*39*)

Interestingly, the effect of the presence or absence of the F_9 segments within the micelle core is similarly reflected in the temperature-dependent ^{1}H NMR behavior of the PPO blocks of CA′BA′C and CABAC as depicted in Figure 8.

Thus, the absence of the perfluoro segments within the micelle core of CABAC (see ^{19}F NMR discussion) resulted in an intense single peak when the PPO–CH_3 and the rm-PGMA–CH_3 signals overlapped at 60 °C.

TFB Uptake-Capabilities of Block Copolymer Micelles

Preliminary experiments show that micelles of BA, CBA, CABAC and CA′BA′C are all capable of solubilizing substantial amount of the perfluorocarbon-based benzene molecule, TFB, as shown in Figure 9. A simple qualitative test at 25 °C showed that the PPO homopolymer is soluble in TFB, hence, the ability of these PPO core bearing micelles to solubilize significant amount of the hydrophobic TFB in aqueous solution. Interestingly, CBA and CA′BA′C micelles showed relatively high solubilization capacities which are approximately 3-4 times higher than that of CABAC and BA micelles. This indicates that the presence of the perfluoro F_9 segments within the micelle cores of CBA and CA′BA′C enhances their solubilization capabilities of the perfluorocarbon-based hydrophobe. This observation confirms the conclusions given by Tuzar and Kratochvil (*40*) that the amount of solubilized compound is mainly controlled by the Flory–Huggins parameter characterizing the enthalpic interaction between the solubilizate and the core-forming copolymer block. Moreover, the fact that CABAC micelles showed solubilization capabilities comparable to that of the BA micelles further confirms that the perfluoro F_9 segments do not form part of the CABAC micelle core.

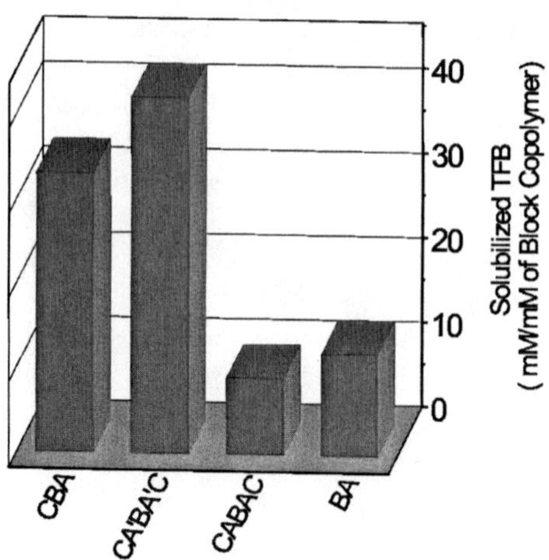

Figure 9. Solubilization of tetrafluorobenzene (TFB) at 25 °C by 5 mg/mL D_2O solutions of the triphilic block copolymers CBA, CA'BA'C, CABAC and the amphiphilic block copolymer BA.

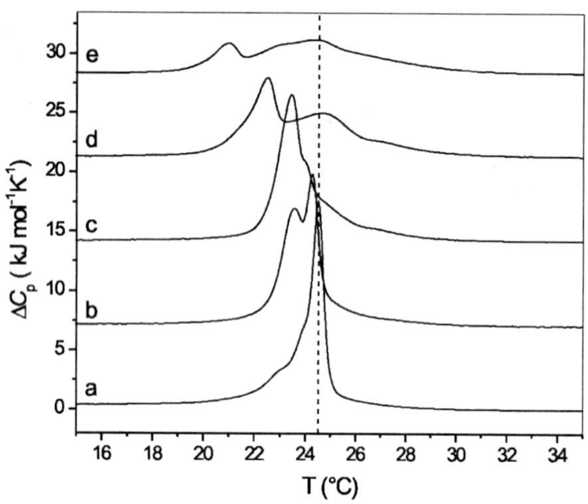

Figure 10. DSC curves in the range of the $L_{\beta'} \rightarrow L_{\alpha}$ phase transition of pure DMPC vesicles (a) and DMPC vesicles in the presence of 1 mol% (b), 2 mol% (c), 5 mol% (d) and 10 mol% (e) CABAC copolymer relative to the lipid concentration. All curves are the 2nd heating traces of a repeated heating/cooling cycle.

Interaction of CABAC Copolymer with Phospholipids Membranes

To check, if the novel triphilic block copolymers interact with lipid membranes we performed DSC measurements on lipid vesicles in the presence of CABAC copolymer. DSC measurements can monitor the gel to liquid-crystalline

($L_{\beta'} \rightarrow L_{\alpha}$) phase transition of lipid membranes. This transition is a cooperative 1st order transition and gives a peak in the C_p= f (T) thermogram, where the integral of the peak corresponds to the phase transition enthalpy and the width is a measure for the cooperativity (the smaller, the more cooperative). The midpoint of the transition is the main transition temperature (T_m) of the lipid membrane. T_m is sensitive to interactions on the membrane surface and/or insertion of hydrophobic substances into the hydrophobic part of the membrane. T_m increases when the interaction leads to a screening of headgroup charges, a replacement of hydration water from the hydrophilic membrane layer (*41*, *42*) or hydrogen bond formation between the headgroups (*43*, *44*). When hydrophobic molecules or moieties insert into the hydrophobic layer of the membrane T_m decreases (*43*, *45*, *46*). The discussed copolymer is triphilic and has the potential to interact with the hydrophilic membrane layer via the PGMA block as well as with the hydrophobic membrane layer via the PPO block (*47*). Also, the perfluoro segment has a tendency to partition into the hydrophobic membrane layer even though it is non-lipophilic. Thus, a shift of T_m would depend on interactions between the lipids and the copolymer and provide an idea about the kind of interaction.

In Figure 10 the thermograms of pure DMPC vesicles and DMPC vesicle suspensions containing different amounts the block copolymer are shown. The copolymer is added to the vesicles after preparation, i.e., the first interaction takes place at the outer monolayer of the vesicles. The lowest curve in Figure 10 shows a typical $L_{\beta'} \rightarrow L_{\alpha}$ transition of DMPC with a T_m value of 24.5 °C (*46*, *48*). The other curves show that the copolymer shifts T_m to lower temperatures. This indicates that the copolymer interacts with the hydrophobic part of the membrane, i.e. the hydrophobic components (PPO and/or F_9) insert into a DMPC membrane. Already 1 mol% of polymer, with respect to the lipids, is sufficient to decrease T_m. The decrease in T_m is more pronounced with increasing copolymer concentration. Between 1 and 10 mol% T_m decreases continuously with increasing copolymer concentration. At 10 mol% polymer concentration, the T_m decrease is about 4.5 °C. In addition, the thermogram does not show a unimodal peak when the copolymer concentration is above 5 mol% but a bimodal peak occurs. One transition component is located at about the same temperature as T_m of a pure DMPC membrane. This component can be attributed to the transition associated with residual unbound DMPC patches or patches with low copolymer content. Thus, the polymer binding leads to phase segregation into ploymer-rich and polymer-poor parts of the membrane. Such phase segregation has already been observed for the binding of polyelectrolytes to charged lipid membranes (*49*, *50*).

The same experiment was repeated with DPPC vesicles. DPPC has the same headgroup as DMPC but its hydrocarbon chains are two CH_2 groups longer, i.e., the hydrophobic membrane layer is thicker. The results are shown in Figure 11. The increased van-der-Waals interactions between the longer hydrophobic chains lead to a higher transition temperature compared to DMPC. T_m of a pure DPPC membrane is about 41.4 °C (*46*, *48*). The presence of the copolymer decreases the T_m of DPPC as already discussed for DMPC membranes. However, much more polymer is required to affect the phase transition of DPPC. At 1 mol% copolymer, with respect to the lipid concentration, T_m of DPPC is not shifted at all. It is only at 10 mol% that a T_m shift to lower temperatures can be seen. But even this shift is small (-0.6 °C) compared to the shift induced by the same polymer concentration to DMPC (-4.5 °C). Moreover, the transition peak of DPPC is even in the presence of the copolymer unimodal. That means, the block copolymer adsorption/insertion does not induce phase segregation to DPPC.

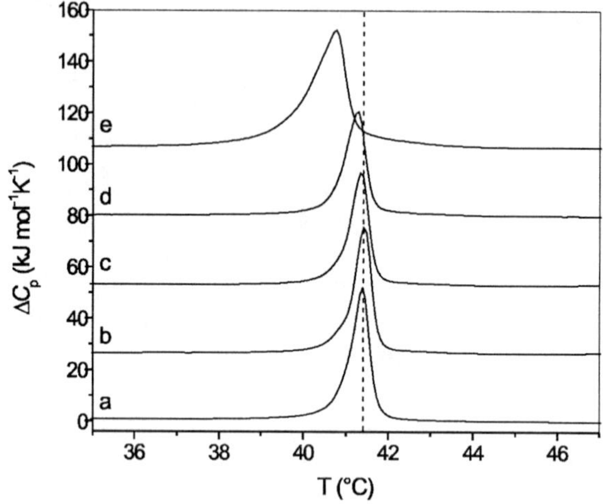

Figure 11. DSC curves in the range of the $L_{\beta'} \rightarrow L_{\alpha}$ phase transition of pure DPPC vesicles (a) and DPPC vesicles in the presence of 1 mol% (b), 2 mol% (c), 5 mol% (d) and 10 mol% (e) CABAC copolymer relative to the lipid concentration. Curves a) to d) represent the 2nd heating trace e) the 3rd heating trace of a repeated heating/cooling cycle.

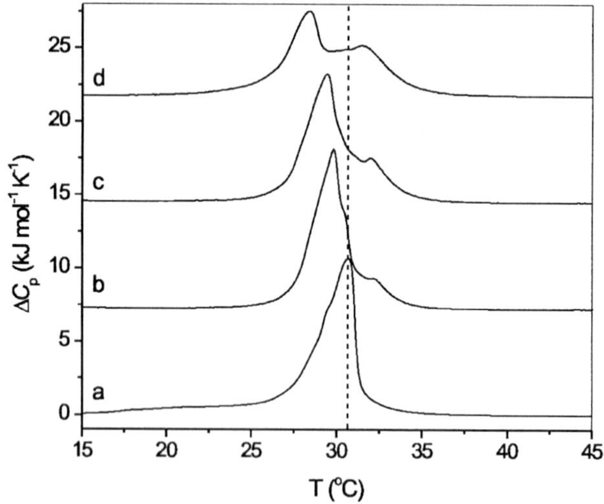

Figure 12. DSC curves in the range of the $L_{\beta'} \rightarrow L_{\alpha}$ phase transition of pure DMPC/DPPC-d_{62} mixed vesicles (a) and DMPC/DPPC-d_{62} mixed vesicles in the presence of 2 mol% (b), 5 mol% (c) and 10 mol% (d) CABAC copolymer relative to the lipid concentration. All curves are the 4th heating traces of a repeated heating/cooling cycle.

This remarkably different interaction between the copolymer and the two homologue lipids leads to the question if the copolymer binding to a mixed DMPC/DPPC membrane can induce lateral phase segregation between the two different lipids within the bilayer. DMPC and DPPC mix ideally which leads to a single cooperative $L_{\beta'} \rightarrow L_{\alpha}$ phase transition in a mixed DMPC/DPPC-d_{62} bilayer at 30.6 °C (*51*) as can be seen in Figure 12. Upon addition of the copolymer this transition splits into two components, indicating phase segregation. One transition component is found below T_m of the undisturbed mixed membrane and decreases continuously in temperature with increasing copolymer concentration. This is very similar to the behavior that is observed for the interactions of pure DMPC with the copolymer. However in presence of 10 mol% copolymer, the transition of this component is only 2.2 °C lowered with respect to the undisturbed mixed membrane, which is less than that observed for pure DMPC/copolymer, but more than observed for the pure DPPC/copolymer system. This indicates that the low temperature peaks represent a phase transition of copolymer bound mixed DMPC/DPPC-d_{62} membrane regions. The second component is observed at temperatures, higher than T_m of the undisturbed mixed bilayer, indicating the melting of a DPPC-d_{62} enriched domain. The position of this transition is independent of the copolymer concentration and is always found at ca. 32 °C. However, its intensity increases with increasing copolymer concentration. From the existence of a DPPC-d_{62} enriched phase it follows that the other phase is depleted in DPPC-d_{62}, i.e., enriched in DMPC. This shows that the copolymer binding induces indeed a lateral demixing between the two homologue lipids, within the temperature range of the phase transition. If this demixing is stable outside the transition range and leads to real phase segregation in the gel and/or the fluid crystalline phase has to be clarified by experiments under variation of the DMPC/DPPC mixing ratio. However, the results clearly show that the copolymer inserts into lipid membranes and that the length of the hydrophobic membrane layer influences the interaction and partitioning.

Conclusions

Novel PPO-based amphiphilic diblock copolymer, N_3-PPO_{27}-$PGMA_{44}$ (BA), and triphilic multiblock copolymer analogues F_9-PPO_{27}-$PGMA_{94}$ (CBA), F_9-$PGMA_{24}$-PPO_{27}-$PGMA_{24}$-F_9 (CABAC), and F_9-$PGMA_{42}$-PPO_{27}-$PGMA_{42}$-F_9 (CA′BA′C) have been successfully synthesized. The aggregation behavior of the copolymers in aqueous solution strongly depends on blocks sequence and length. Because they are covalently linked to each other, both the F_9 segment and the PPO block formed the core structure of CBA micelles. Also, for CA′BA′C micelles both the F_9 segment and the PPO block formed the core structure via looping of the PGMA blocks. However, for CABAC the F_9 segment forms part of the micelle corona, hence, it is absent in the micelle core. The lack of the F_9 segments within the core of CABAC micelles is due to the inability of their short rigid PGMA blocks to loop. The effect of the presence or the absence of the F_9 segment within the micelle cores of the block copolymers was reflected in the temperature-dependent ^1H NMR behavior of their PPO blocks. Interestingly, while CBA and CA′BA′C micelle cores showed high tetrafluorobenzene uptake-capabilities, due to the presence of the fluorophilic F_9 segments, CABAC miclle core showed uptake-capabilities similar to BA. This further confirmed the absence of the F_9 segments within the micelle cores of CABAC. It was found that CABAC inserts into lipid membranes thereby drecreasing the the gel to liquid-crystalline ($L_{\beta'} \rightarrow L_{\alpha}$) phase transition tempearture of the lipid membranes. The effect is more pronounced in the membranes of DMPC than DPPC and leads to phase segregation into ploymer-rich and polymer-poor parts above 5 mol% polymer concentration in DMPC. Futhermore, CABAC induces lateral demixing between the DMPC/DPPC-d_{62} lipids in the temperature range of the phase transition of the lipids. First measurements indicate that block copolymers having perfluoro units are able to influence the uptake of doxorubicin by multiple drug resistant cancer cells (*52*).

Acknowledgments

We thank the Deutsche Forschungsgemeinschaft (DFG), FOR 1145 for the financial support.

References

1. Amado, E.; Blume, A.; Kressler, J *React. Funct. Polym.* **2009**, *69*, 450–456.
2. Chu, B.; Zukang, Z. In *Nonionic Surfactants, Polyoxyalkylene Block Copolymers*; Nace, V. M., Ed.; Marcel Dekker, Inc.: New York, 1996; pp 67−144.
3. Munshi, N.; Rapoport, N.; Pitt, W. G. *Cancer* **1997**, *118*, 13–19.
4. Chiappetta, D. A.; Sosnik, A. *Eur. J. Pharm. Biopharm.* **2007**, *66*, 303–317.
5. Escobar-Chavez, J. J.; Lopez-Cervantes, M.; Naik, A.; Kalia, Y. N.; Quintana-Guerrero, D.; Ganem-Quintanar, A. *J. Pharm. Pharm. Sci.* **2006**, *3*, 339–358.
6. Kabanov, A. V.; Batrakova, E. V.; Alakhov, V. Y. *J. Controlled Release* **2002**, *82*, 189–212.
7. Melik-Nubarov, N. S.; Pomaz, O. O.; Dorodnych, T.; Badun, G. A.; Ksenofontov, A. L.; Schemchukova, O. B.; Arzhakov, S. A. *FEBS Lett.* **1999**, *446*, 194–198.
8. Amado, E.; Kerth, A.; Blume, A.; Kressler, J. *Langmuir* **2008**, *24*, 10041–10053.
9. Mequanint, K.; Patel, A.; Bezuidenhout, D. *Biomacromolecules* **2006**, *7*, 883–891.
10. Wang, C.; Yu, B.; Knudsen, B.; Harmon, J.; Moussy, F.; Moussy, Y. *Biomacromolecules* **2008**, *9*, 561–567.
11. Gates, G.; Harmon, J. P.; Ors, J.; Benz, P. *Polymer* **2003**, *44*, 207–214.
12. Mao, J.; Gan, Z. *Macromol. Biosci.* **2009**, *9*, 1080–9.
13. Halacheva, S.; Rangelov, S.; Tsvetanov, C. *Macromolecules* **2006**, *39*, 6845–6852.

14. Kyeremateng, S. O.; Amado, E.; Blume, A.; Kressler, J. *Macromol. Rapid Commun.* **2008**, *29*, 1140–1146.
15. Sumerlin, B. S.; Vogt, A. P. *Macromolecules* **2010**, *43*, 1–13.
16. Lodge, T. P.; Rasdal, A.; Li, Z.; Hillmyer, M. A. *J. Am. Chem. Soc.* **2005**, *127*, 17608–17609.
17. Gerebtzoff, G.; Li-Blatter, X.; Fischer, H.; Frentzel, A.; Seelig, A. *ChemBioChem* **2004**, *5*, 676–684.
18. Kyeremateng, S. O.; Amado, E.; Kressler, J. *Eur. Polym. J.* **2007**, *43*, 3380–3391.
19. Lutz, J.-F.; Matyjaszewki, K. *J. Polym. Sci., Part A: Polym. Chem.* **2005**, *43*, 897–910.
20. Coessens, V.; Nakagawa, Y.; Matyjaszewski, K. *Polym. Bull.* **1998**, *40*, 135–142.
21. Binder, W. H.; Sachsenhofer, R.; Straif, C. J.; Zirbs, R. *J. Mater. Chem.* **2007**, *17*, 2125–2132.
22. Altinok, H.; Nixon, S. K.; Gorry, P. A.; Attwood, D.; Booth, C.; Kelarakis, A.; Havredaki, V. *Colloids Surf., B* **1999**, *16*, 73–91.
23. Lopes, J. R.; Loh, W. *Langmuir* **1998**, *14*, 750–756.
24. Preuschen, J.; Menchen, S.; Winnik, M. A.; Heuer, A.; Spiess, H. W. *Macromolecules* **1999**, *32*, 2690–2695.
25. Zhang, H.; Pan, J.; Hogen-Esch, T. E. *Macromolecules* **1998**, *31*, 2815–2821.
26. Kyeremateng, S. O.; Henze, T.; Busse, K.; Kressler, J. *Macromolecules* **2010**, *43*, 2502–2511.
27. Carriere, P.; Grohens, Y.; Spevacek, J.; Schultz, J. *Langmuir* **2000**, *16*, 5051–5053.
28. Chen, G.; Zhu, X.; Cheng, Z.; Lu, J.; Chen, J. *Polym. Int.* **2004**, *53*, 357–363.
29. Wang, J.-S.; Matyjaszewski, K. *Macromolecules* **1995**, *28*, 7901–7910.
30. Cau, F.; Lacelle, S. *Macromolecules* **1996**, *29*, 170.
31. Gottlieb, H. E.; Kotlyar, V.; Nudelman, A. *J. Org. Chem.* **1997**, *62*, 7512–7515.
32. Ma, J.-h.; Guo, C.; Tang, Y.-l.; Liu, H.-z. *Langmuir* **2007**, *23*, 9596–9605.
33. Guo, C.; Liu, H. Z.; Chen, J. Y. *Colloid Polym. Sci.* **1999**, *277*, 376–381.
34. Hvidt, S.; Trandum, C.; Batsberg, W. *J. Colloid Interface Sci.* **2002**, *250*, 243–250.
35. Goldmints, I.; Yu, G.-E.; Booth, C.; Smith, K. A.; Hatton, T. A. *Langmuir* **1999**, *15*, 1651–1656.
36. Almgren, M.; Brown, W.; Hvidt, S. *Colloid Polym. Sci.* **1995**, *273*, 2–15.
37. Yang, L.; Alexandridis, P.; Steyler, D. C.; Kositza, M. J.; Holzwarth, J. F. *Langmuir* **2000**, *16*, 8555–8561.
38. Pedersen, J. S.; Gerstenberg, M. C. *Colloids Surf., A* **2003**, *213*, 175–187.
39. Save, M; Weaver, J. V. M.; Armes, S. P.; McKenna, P. *Macromolecules* **2002**, *35*, 1152–1159.
40. Tuzar, Z.; Kratochvil, P. In *Surface and Colloid Science*; Matijevic, E., Ed.; Plenum Press: New York, 1993; pp 1–83.
41. Cevc, G.; Watts, A.; Marsh, D. *FEBS Lett.* **1980**, *120*, 267–270.
42. Watts, A.; Harlos, K.; Maschke, W.; Marsh, D. *Biochem. Biophys. Acta* **1978**, *510*, 63–74.

43. Papahadjopoulos, D.; Moscarello, M.; Eylar, E. H.; Isac, T. *Biochem. Biophys. Acta* **1975**, *401*, 317–335.
44. Schwieger, C.; Blume, A. *Eur. Biophys. J.* **2007**, *36*, 437–450.
45. Ivanova, V. P.; Makarov, I. M.; Schaffer, T. E.; Heimburg, T. *Biophys. J.* **2003**, *84*, 2427–2439.
46. Huang, C. H.; Li, S. *Biochem. Biophys. Acta* **1999**, *1422*, 273–307.
47. Amado, E.; Kerth, A.; Blume, A.; Kressler, J. *Soft Matter* **2009**, *5*, 669–675.
48. Blume, A. *Thermochim. Acta* **1991**, *193*, 299–347.
49. Schwieger, C.; Blume, A. *Biomacromolecules* **2009**, *10*, 2152–2161.
50. Schwieger, C.; Blume, A. *Eur. Biophys. J.* **2007**, *36*, 437–450.
51. Blume, A.; Garidel, P. In *Handbook of Thermal Analysis Calorimetry*; Kemp, R. B., Ed.; Elsevier Science B.V.: Amsterdam, 1999; Vol. 4, pp 109–173.
52. Sommer, K.; Kaiser, S.; Krylova, O. O.; Kressler, J.; Pohl, P.; Busse, K. *J. Med. Chem.* **2008**, *51*, 4253–4259.

Chapter 6

Polymer Blends from Optoelectronics to Spintronics

Liang Yan, Yue Wu, and Bin Hu*

Department of Materials Science and Engineering, University of Tennessee, Knoxville, Tennessee 37996, USA
***bhu@utk.edu**

This chapter reports recent experimental studies on electro-optically active polymer blends in organic spintronics. The experimental results indicate that polymer blends offer a convenient methodology to modify the critical parameter: spin-orbital coupling, in spintronics through inter-molecular interaction. Furthermore, the energy transfer in polymer blends can carry magnetic field effects from one component to another and consequently amplify the magnetic field effects in polymer blends. As a result, polymer blends are an important class of materials in organic spintronics.

Introduction

Polymer blending presents a fundamental method to generate nanoscale morphological structures. The nanoscale morphological structures can offer effective control on charge transport and excited processes in optoelectronics where electronic and optic processes can be mutually controlled. Recently, experimental studies have found that polymer blends can have tunable inter-molecular spin-orbital coupling which is a critical parameter in spintronics where magnetic, optic, and electronic processes can be mutually controlled. As a result, polymer blends have become an important class of materials from optoelectronics to spintronics. In optoelectronics organic light emitting diodes (OLEDs) have been widely investigated due to their high potential applications in flexible display and large-area solid-state lighting (*1*, *2*). Polymer molecular composites (*3*) or polymer blends (*4*) are widely used to control the key processes: balancing degree of bipolar electron and hole injection, electron-hole recombination, energy transfer, and light emission efficiency in polymer based white OLEDs. Furthermore, it has recently been found from magnetic field effect on electroluminescence (MFE_{EL}) that the light intensity changes with magnetic field in the OLEDs (*5–7*). This experimental finding indicates that organic semiconducting materials including polymers and polymer blends can be used for organic spintronics. There are two possible mechanisms for MFE_{EL}. One is that the magnetic field changes the formation rate of both the singlet and triplet excited states and further leads to their emission changing (*8*). The other possible mechanism is that the external magnetic field changes singlet/triplet excited states ratio by affecting intersystem crossing (*6, 9, 10*). In the intersystem crossing mechanism, the competition between internal magnetic interaction, such as spin-orbital coupling, and the external magnetic field is very important to determine singlet/triplet ratio change caused by the external magnetic field in organic semiconductors. Polymer molecular composites or polymer blends can also show MFE_{EL}. In these mixed structures, energy transfer and spin-orbital coupling are two existing key factors in the determination of magnetic field effects. In our recent work, we studied the MFE_{EL} in polymer blend-based devices with energy transfer, in composites of poly(N-vinylcarbazole) (PVK) and tris[2-phenylpyridine] iridium ($Ir(ppy)_3$). We observed by using energy transfer from a strong spin-orbital-coupling material to a weak spin-orbital-coupling material that the MFE_{EL} in strong spin-orbital coupling materials could be highly amplified. This experimental finding indicates that the use of polymer blends presents a new opportunity to amplify magnetic field effects in organic spintronics.

Experimental

The polymer molecular blend was prepared by dispersing the heavy-metal complex $Ir(ppy)_3$ into a poly(N-vinylcarbazole) (PVK) matrix at 0.1wt%. The PVK and $Ir(ppy)_3$ blend was dissolved in chloroform and spun cast on pre-cleaned indium tin oxide (ITO) coated glass substrates. An aluminum metal electrode was then thermally evaporated on the PVK+$Ir(ppy)_3$ blend film under a vacuum of 2×10^{-6} Torr to fabricate light emitting devices with the architecture of ITO/PVK+$Ir(ppy)_3$/Al. The MFE_{EL} was measured at constant current density 20 mA/cm^2 in liquid nitrogen. The MFE_{EL} was defined as $\frac{I_{EL(B)} - I_{EL(0)}}{I_{EL(0)}}$, where $I_{EL(B)}$ and $I_{EL(0)}$ are the electroluminescence intensity at constant current condition with and without magnetic field, respectively.

Result and Discussion

It is known that the electroluminescence is generated by radiative emission of intra-molecular electron-hole pairs, namely excitons. We know that the excitons have both singlet (S) and triplet (T) states. In general, the singlet/triplet ratio in excitonic states is 1/3 when electron-hole pairs are formed. A magnetic field can change electroluminescence by changing the singlet and triplet formation rate or by changing singlet-triplet intersystem crossing. For singlet and triplet formation, the experimental result that both of fluorescence and phosphorescence emission in electroluminescence increase with applied magnetic field suggests that both singlet and triplet exciton formation rates increase under the influence of a magnetic field (*8*). For singlet-triplet intersystem crossing, an external magnetic field needs to be strong enough, relative to the internal magnetic field generated by spin orbital coupling, to generate MFE_{EL}. When strong spin-orbital-coupling $Ir(ppy)_3$ molecules are dispersed in a weak spin-orbital-coupling PVK matrix, the penetration of delocalized π electrons from the PVK matrix into the magnetic field from an orbital current in the $Ir(ppy)_3$ molecules can inevitably generate inter-molecular spin-orbital coupling. This inter-molecular spin-orbital coupling can be largely tuned by adjusting the $Ir(ppy)_3$ dispersion concentration. In particular, this tunable inter-molecular spin-orbital coupling forms a convenient and effective mechanism to modify the spin-orbital coupling strengths for each component in such polymer blends.

Figure 1a shows the MFE_{EL} from individual PVK and pure $Ir(ppy)_3$ components. The characteristic electroluminescence peaks are around 405 nm and 505 nm for individual PVK and $Ir(ppy)_3$ components. It should be noted that the pure PVK exhibits a large MFE_{EL} with an amplitude of about +10% but the pure $Ir(ppy)_3$ does not show an appreciable MFE_{EL}. It is obvious that the PVK yields fluorescence from its singlet excitons while the $Ir(ppy)_3$ generates phosphorescence from its triplet excitons due to the strong spin-orbital coupling caused by the heavy metal Ir complex structure. For weak spin-orbital-coupling materials such as PVK, a low magnetic field of around 10~100 mT is stronger than the internal magnetic field from spin-orbital coupling. As a consequence, significant MFE_{EL} can be expected. For the strong-spin-orbital-coupling materials such as $Ir(ppy)_3$, a low magnetic field less than 1T is much weaker than the internal magnetic field generated by spin-orbital coupling (*11*). As a result, negligible MFE_{EL} can be observed by a first order approximation from the pure $Ir(ppy)_3$ electroluminescence.

After we lightly doped $Ir(ppy)_3$ into the PVK matrix, the PVK + $Ir(ppy)_3$ blend spectrum (Figure 2a) shows both characteristic peaks at 405 nm from the PVK matrix and at 505 nm from the dispersed $Ir(ppy)_3$ molecules in electroluminescence. More importantly, both electroluminescence peaks show clear MFE_{EL}. As shown in Figure 2b and 2c, the PVK matrix is around 5% MFE_{EL} smaller than the pure PVK device. But the dispersed $Ir(ppy)_3$ surprisingly shows +3% MFE_{EL} while the pure $Ir(ppy)_3$ device gives negligible MFE_{EL}. The decrease of MFE_{EL} for the PVK matrix compared to pure PVK is due to the increase in its spin-orbital coupling strength caused by the inter-molecular spin-orbital coupling (*12*). The positive MFE_{EL} for the dispersed $Ir(ppy)_3$ molecules can come from two possibilities. First, the spin-orbital coupling strength of dispersed $Ir(ppy)_3$ molecules was weakened by the PVK matrix through inter-molecular spin-orbital coupling effects. Then a low magnetic field becomes strong enough to compete with the weakened spin-orbital coupling strength in the dispersed $Ir(ppy)_3$. Thus the electroluminescence from the dispersed $Ir(ppy)_3$ molecules can exhibit significant MFE_{EL}. Second, the energy transfer from the PVK matrix to dispersed Ir(py)$_3$ molecules can carry the MFE_{EL} from the PVK matrix to the dispersed Ir(ppy)$_3$ molecules, leading to a MFE_{EL} in the dispersed Ir(ppy)$_3$ in the PVK + Ir(ppy)$_3$ blend. The observed positive MFE_{EL} from the dispersed Ir(ppy)$_3$ molecules can rule out the first possibility a spin-orbital coupling mechanism, because the spin-orbital coupling mechanism should give negative MFE_{EL} in triplet electroluminescence through intersystem crossing (*10*). Therefore, the energy transfer mechanism is mainly accountable for the positive MFE_{EL} observed from the PVK + Ir(ppy)$_3$ blend.

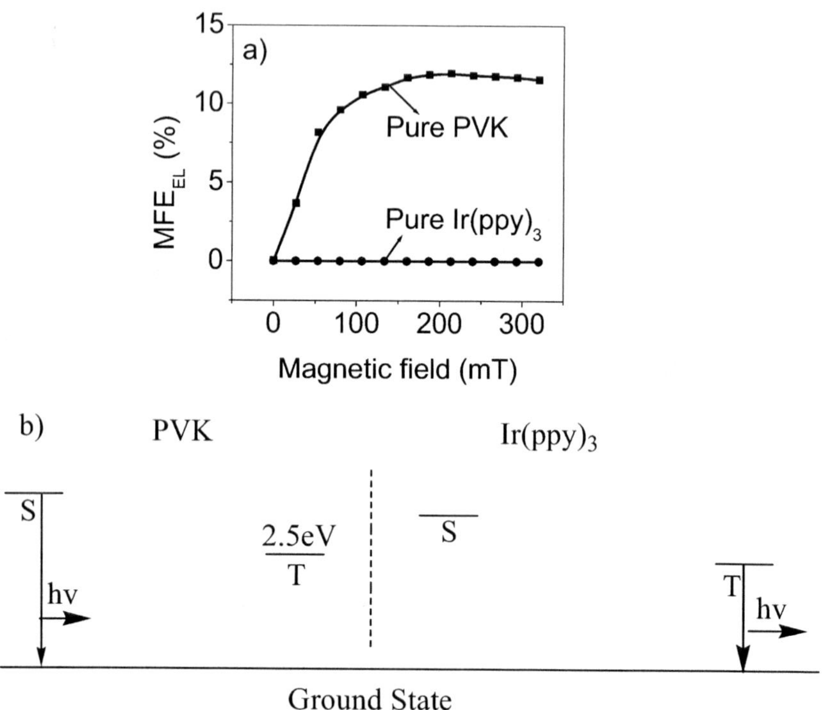

Figure 1. a) MFE_{EL} for pure PVK and Ir(ppy)$_3$, b) The singlet and triplet energy levels for both PVK and Ir(ppy)$_3$. S and T refer to singlet and triplet states.

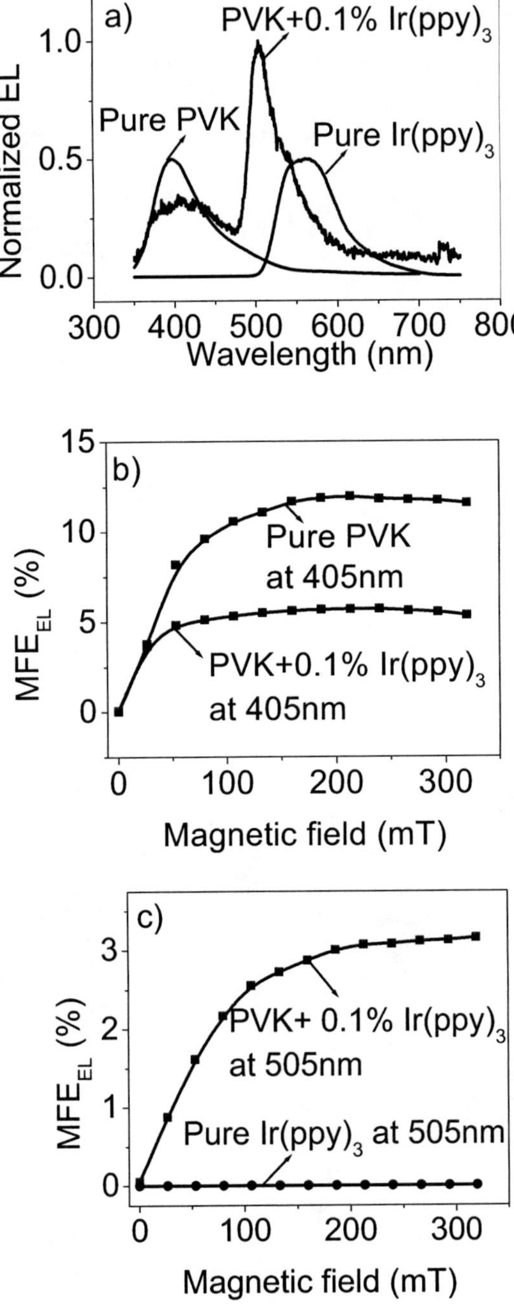

Figure 2. a) Normalized electroluminescence spectrum for pure PVK, pure $Ir(ppy)_3$ and PVK+0.1% $Ir(ppy)_3$ blend, b) MFE_{EL} at 405 nm for 0.1% $Ir(ppy)_3$+PVK blend and pure PVK, c) MFE_{EL} at 505nm for 0.1% $Ir(ppy)_3$+PVK blend and pure $Ir(ppy)_3$.

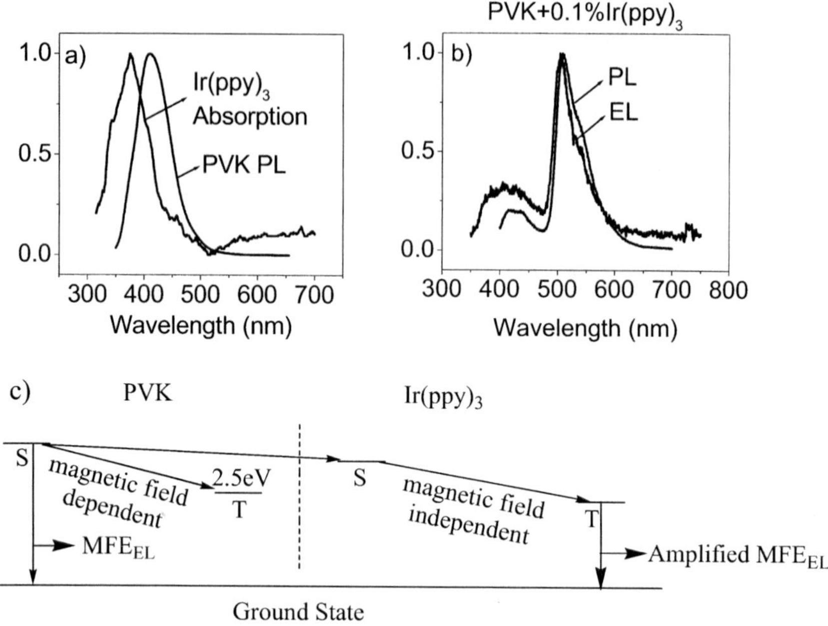

Figure 3. a) Absorption of Ir(ppy)3 and photoluminescence of PVK, b) Photoluminescence and electroluminescence of PVK with 0.1% Ir(ppy)3 , c) Energy transfer process in electroluminescence of composite of PVK doped with 0.1wt% Ir(ppy)3. S and T refer to singlet and triplet states.

Figure 3a shows the absorption spectrum of $Ir(ppy)_3$ and photoluminescence spectrum of PVK. We can see that the PVK emission and $Ir(ppy)_3$ absorption have a large spectral overlap to activate the Förster energy transfer from the PVK matrix to the dispersed $Ir(ppr)_3$ molecules. Figure 3b shows both photoluminescence and electroluminescence spectra for the PVK + 0.1% $Ir(ppy)_3$ blend. It should particularly be noted that the photoluminescence and electroluminescence spectra are quite similar. Both of them show two peaks: one from PVK at 405nm, one from $Ir(ppy)_3$ at 505nm. The spectral similarity between photoluminescence and electroluminescence implies that the Förster energy transfer occurs in the PVK+$Ir(ppy)_3$ blend device. Furthermore, the significant electroluminescence quenching in PVK matrix caused by the low 0.1% $Ir(ppy)_3$ doping indicates that the Förster energy transfer is very efficient in the PVK + $Ir(ppy)_3$ blend device.

Figure 3c schematically illustrates how energy transfer can amplify the MFE_{EL} for the phosphorescence of heavy-metal complex $Ir(ppy)_3$ through polymer blend design. When PVK and $Ir(ppy)_3$ were mixed, a low magnetic field would change the intersystem crossing in PVK matrix with the consequence of increasing singlet ratio in the PVK matrix. Then due to the energy transfer, the increased singlets in the PVK matrix are transferred to the singlet states in the dispersed $Ir(ppy)_3$ molecules through the Förster process. Because of the strong spin-orbital coupling in the dispersed $Ir(ppy)_3$ molecules, the intersystem crossing process can be very efficient and especially independent on magnetic field in the Ir(ppy)$_3$ component. As a result, almost all singlets are converted into triplets in the dispersed Ir(ppy)$_3$ molecules. This means that triplet concentration in the dispersed Ir(ppy)$_3$ molecules can essentially increase due to (i) efficient energy transfer from the PVK matrix to the dispersed Ir(ppy)$_3$ molecules and (ii) strong intersystem crossing within the dispersed Ir(ppy)$_3$ molecules. Therefore, the MFE_{EL} from the dispersed Ir(ppy)$_3$ molecules can be considerably amplified based on polymer blend design as compared to pure Ir(ppy)$_3$ molecules.

Conclusion

We have shown that polymer blend design can lead to a substantial tuning of spin-orbital coupling through inter-molecular interaction when a weak spin-orbital-coupling polymer is mixed with a strong spin-orbital-coupling molecules. The tuning of spin-orbital coupling comes from the penetration of delocalized π electrons from a weak spin-orbital-coupling polymer matrix into the magnetic field of orbital current from strong spin-orbital-coupling molecules. On the other hand, the Förster energy transfer can occur from the polymer matrix to dispersed molecules in polymer molecular blends. Especially, Förster energy transfer can carry the MFE_{EL} from weak-spin-orbital-coupling polymer matrix to strong spin-orbital-coupling molecules, leading to amplification on the MFE_{EL} in the strong spin-orbital-coupling molecules. Therefore, polymer blends design presents a new mechanism to amplify magnetic field effects in organic spintronics based on (i) modification of spin-orbital coupling and (ii) Förster energy transfer.

References

1. Kanno, H.; Sun, Y.; Forrest, S. R. High-efficiency top-emissive white-light-emitting organic electrophosphorescent devices. *Appl. Phys. Lett.* **2005**, *86*, 263502–263503.
2. Tokito, S.; Iijima, T.; Tsuzuki, T.; Sato, F. High-efficiency white phosphorescent organic light-emitting devices with greenish-blue and red-emitting layers. *Appl. Phys. Lett.* **2003**, *83*, 2459–2461.
3. Al Attar, H. A.; Monkman, A. P.; Tavasli, M.; Bettington, S.; Bryce, M. R. White polymeric light-emitting diode based on a fluorene polymer/Ir complex blend system. *Appl. Phys. Lett.* **2005**, *86*.
4. Hu, B.; Karasz, F. E. Blue, green, red, and white electroluminescence from multichromophore polymer blends. *J. Appl. Phys.* **2003**, *93*, 1995–2001.
5. Kalinowski, J.; Cocchi, M.; Virgili, D.; Macro, P. D.; Fattori, V. Magnetic field effects on emission and current in Alq3-based electroluminescent diodes. *Chem. Phys. Lett.* **2003**, *380*, 710.
6. Kalinowski, J.; Cocchi, M.; Virgili, D.; Fattori, V.; Di Marco, P. Magnetic field effects on organic electrophosphorescence. *Phys. Rev. B* **2004**, *70* (7), 205303.
7. Davis, A. H.; Bussmann, K. Large magnetic field effects in organic light emitting diodesbased on tris(8-hydroxyquinoline aluminum)

(Alq$_3$)/N,N′-Di(naphthalene-1-yl)-N, N′ diphenyl-benzidine (NPB) bilayers. *J. Vac. Sci. Technol., A* **2004**, *22*, 1885.

8. Reufer, M.; et al. Spin-conserving carrier recombination in conjugated polymers. *Nat. Mater.* **2005**, *4*, 340–346.
9. Hu, B.; Wu, Y. Tuning magnetoresistance between positive and negative values in organic semiconductors. *Nat. Mater.* **2007**, *6*, 985–991.
10. Hu, B.; Yan, L.; Shao, M. Magnetic-field effects in organic semiconducting materials and devices. *Adv. Mater.* **2009**, *21*, 1500.
11. Birks, J. B. *Organic Molecular Photophysics*; Wiley: London, 1975.
12. Wu, Y.; Xu, Z.; Hu, B.; Howe, J. Tuning magnetoresistance and magnetic-field-dependent electroluminescence through mixing a strong-spin-orbital-coupling molecule and a weak-spin-orbital-coupling polymer. *Phys. Rev. B* **2007**, *75*, 035214–035216.

Chapter 7

Preparation and Characterization of Hydrogels of Several Polysaccarides for Biomaterials Applications

Hydrogels for Biomaterials Applications

Ayse Z. Aroguz,[*,1] Kemal Baysal,[2] and Bahattin M. Baysal[3]

[1]Department of Chemistry, Engineering Faculty, İstanbul University, 34320 Avcılar-Istanbul, Turkey
[2]TUBITAK Research Institute for Genetic Engineering and Biotechnology, Marmara Research Center, P.O. Box 21, 41470 Gebze-Kocaeli, Turkey
[3]Department of Chemical Engineering, Bogazici University, 81815 Bebek-Istanbul, Turkey
[*]aroguz@istanbul.edu.tr

The advantages of hydrogels as biomedical materials and their performance in applications depend on their molecular structures. Two hydrogels derived from biopolymers such as alginate and chitosan obtained from natural sources were prepared and characterized for biomaterials applications. Morphology of the crosslinked material was investigated by using SEM. Swelling behavior of chitosan/polycaprolactone and chitosan/alginate hydrogels in different compositions was studied. The biopolymer hydrogels thus prepared were used for cell growth experiments. To measure the degree of cell proliferation, the gels were seeded with L929 mouse fibroblasts. The cell growth properties were followed by SEM photographs. It was found that the population of cells in the cross-linked polymer gels increased.

Introduction

The structure and properties of polymeric gels are important for their applications in tissue engineering (*1–4*).

Chitosan is linear polysaccaride commenly used in the tissue engineering applications for its biocompability (*5–7*). Chitosan is often crosslinked with other polymers such as polycaprolactone and alginate to modulate its general properties such as mechanical property for biomedical applications (*8*).

Alginate is another natural polysaccaride obtained from brown seaweeds. It is commonly used in the food industry as additives and thickeners (*9*). It has also an excellent biocompatibility, potential bioactivity and non-toxicity and is used in tissue engineering drug systems (*10*, *11*).

Many review articles have been published on the use of natural polymeric materials in a broad range of biomedical applications (*12–14*)

Polycaprolactone (PCL), an aliphatic polyester, is one of the most important synthetic materials for commercial applications as a biodegradable material. Recently, blending PCL with the other polymers has been studied to obtain a more economical version of the degradable polymer (*15*, *16*). Blends of polycaprolactone with chitosan were studied for tissue engineering applications by Madihally and coworkers (*17*).

Our laboratories have extensively studied preparation, characterization and various properties and applications of these biomaterials in drug release and cell growth potentials (*18–22*).

In this study, chitosan/alginate and chitosan/polycaprolactone crosslinked gels were prepared in the presence of the cross-linking agent. To obtain a porous polymeric structure, solvent removal method by freeze-drying was used. The swelling behavior and morphological properties of the gels were studied. Adding acrylic acid (AA) to the chitosan/poly caprolactone crosslinked material during the preparation process increased the stability of the hydrogels.

Experimental

Materials

Chitosan powder with a medium molecular weight, alginic acid sodium salt and polycaprolactone with the average molecular weight (Mw) of 80,000 g/mol were supplied from Aldrich Chemical Company Inc, U.S.A. Acrylic acid was purchased from Fluka. The crosslinker, N,N′- dicyclohexylcarbodiimide (DCC) was also purchased from Aldrich Cartsan DBTDL, stannous dilaurate (T-12) used as catalyst was obtained from Cincinnati Milacron Chemicals Inc., U.S.A. All the polymers were used without further purification.

Preperation of Hydrogels

The gel preparation reactions were performed in acetic acid solutions to obtain products in high yields. First, stock solutions of the polymers were prepared: Chitosan was dissolved in 0.5 M acetic acid at 121 °C; Stock solutions of alginate and polycaprolactone were prepared in water and in glacial acid at room temperature, respectively. The polycaprolactone solution was slowly added to the chitosan solution under continuous stirring.

To obtain crosslinked chitosan/PCL gel, catalyst T-12 and DCC crosslinker were added into the mixture. The crosslinked product was poured into Teflon petri dishes of 2 cm diameter, and subsequently evaporated in a hood at room temperature until a concentrated gel was obtained. All samples were prepared by using the same procedure, but at different mixing ratios of the original materials.

The same procedure was also used to prepare chitosan/alginate crosslinked gels. The alginate solution was slowly added to the chitosan solution, then T-12 and DCC were added to this mixture to obtain a crosslinked chitosan/alginate hydrogel.

Chitosan/PCL/AA gels were prepared using the same procedure as the one for the chitosan/PCL hydrogel. AA was added to the mixture before adding the catalyst and crosslinker. All the samples were evaporated in the Teflon petri dishes to obtain concentrated hydrogels.

Lyophilization

Lyphlock 6, a Labconco Model Freeze Drier system, was used for lyophlisation of the crosslinked materials and obtaining porous structures. The concentrated gel products were first frozen for 2 hours at −80°C. Then the frozen gels were lyophilized within the freeze drier for 6 hours at –50°C to obtain the porous scaffold. The lyophilized freeze-dried samples were used in the cell growth and drug release experiments.

Swelling of Gels

Dynamic swelling experiments of the hydrogels were performed by immersing dried disc samples in distilled water (pH=5.6) and in phosphate buffered saline solution (PBS) (pH=7.4) in a temperature-controlled bath at 298 ± 0.1 K. The mass of the gel sample was messured as a function of time. The equilibrium mass swelling is expressed as the amount of water in the hydrogel:

$$\text{Equilibrium mass swelling} = (m_{\infty}-m_o)/m_o \qquad (1)$$

Where, m_0 is the mass of the dry gel and $m\infty$ the mass of the swollen gel at equilibrium. A sample of the chitosan/PCL gel was dried in vacuum oven at 50 °C until a constant weight of the gel was reached. Then certain amount of chitosan/PCL dry gel film was immersed in distilled water and in PBS separately, and stored at room temperature for several hours. The gels were removed from the solution and wiped gently with tissue paper and weighted quickly. Thus, the hdyrogel was weighed in the swollen state, and the equilibrium mass swelling was calculated using Equation (1).

The equilibrium mass swelling values were compared for various compositions of the gel and different amounts of the crosslinker used in the gel.

Morphological Analysis

The morphology of the crosslinked gels was analyzed by a JOEL-JSM T330 scanning electron microscope (SEM). Ambios Universal Scanning Probe Microscope AFM was also used to compare the morphologies of the samples. For SEM observation, the samples were coated with gold before analysis. The coated samples were placed in the vacuum chamber of the instrument and observed at the voltage of 5 kV. The results are shown in Figs. 1 for a chitosan / PCL gel and Fig. 2-3 for a Chitosan/alginate gel. Figs. 4-5 show the AFM results of the chitosan/alginate hydrogel.

Results and Discussion

Table 1. The composition and swelling properties of the Chitosan/PCL hydrogel

<table>
<tr><th>Chitosan/PCL (w/w)</th><th>Swelling medium</th><th>Crosslinker (wt%) in the original mixture</th><th>Equilibrium mass swelling ratio</th></tr>
<tr><td rowspan="3">67/33</td><td rowspan="3">water</td><td>0.83</td><td>40</td></tr>
<tr><td>0.67</td><td>91</td></tr>
<tr><td>0.41</td><td>180</td></tr>
<tr><td rowspan="2">50/50</td><td>water</td><td rowspan="2">0.70</td><td>95</td></tr>
<tr><td>PBS</td><td>48</td></tr>
<tr><td>25/75</td><td rowspan="3">water</td><td rowspan="3">1.5</td><td>17</td></tr>
<tr><td>50/50</td><td>53</td></tr>
<tr><td>75/25</td><td>58</td></tr>
<tr><td>25/75</td><td rowspan="3">PBS</td><td rowspan="3">1.5</td><td>16</td></tr>
<tr><td>50/50</td><td>32</td></tr>
<tr><td>75/25</td><td>42</td></tr>
<tr><td rowspan="2">Chitosan/PCL/AA (w/w/w) 50/25/25</td><td>water</td><td rowspan="2">0.37</td><td>14</td></tr>
<tr><td>PBS</td><td>6</td></tr>
</table>

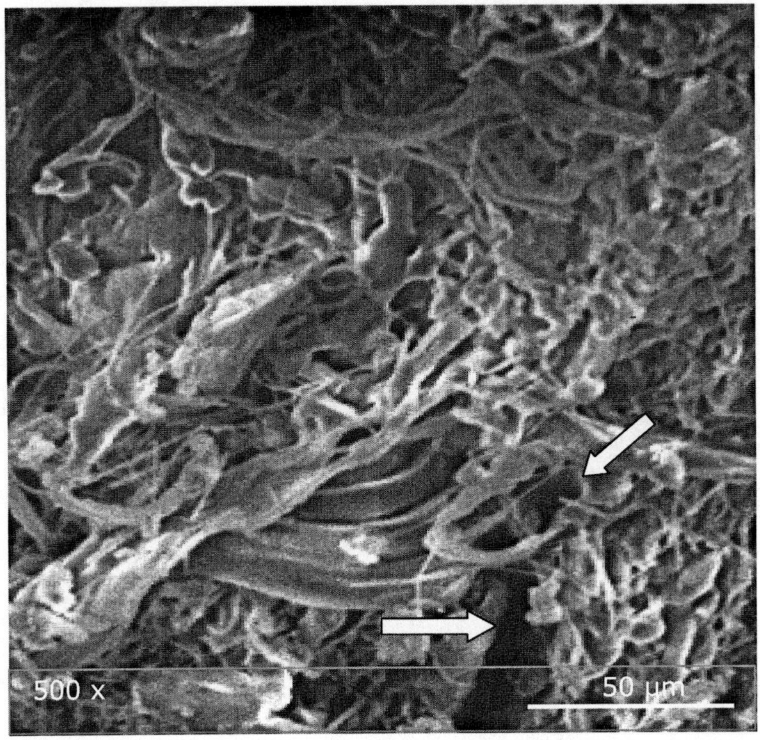

Figure 1. SEM micrograph of the surface morphology of chitosan/poly caprolactone (1/1) (x500).

Swelling Ratio

The equilibrium mass swelling ratios of the chitosan/PCl and chitosan/ alginate hydrogels are listed in Table 1 and Table 2, respectively.

In Table 1 the mass swelling ratio increases with a decreasing crosslinker content in the gel. The ratio also increases with an increasing amount of chitosan in the gel. Similar results are observed for the chitosan/alginate hydrogels (Table 2). An increasing amount of chitosan in the chitosan/alginate gel increases the swelling ratio. The mass swelling in PBS solution is less than it is in water for the two hydrogels. Adding AA decreases the swelling of the two gels.

Morphology

Figure 1 shows the morphological structure of the chitosan/alginate hydrogel. A porous and fibrous sructure is evident. The crystalline structure between the holes comes from PCL. The arrows in Figure 1 represent the holes. Figure 2 shows the surface of the chitosan /alginate hydrogel. The surface of the chitosan /alginate hydrogel after being seeded with cells is seen in Figure 3. The arrows in Figure 3 show the growing seeded cell. AFM results of chitosan/alginate hydrogel are seen in Figs. 4-5. AFM images support the porous structures of the gels. These images reveal a homogenous distrubution of holes in the gels. The porous structure is suitable for use as a scaffold for cell growth experiments.

Table 2. The composition and swelling properties of the Chitosan/alginate hydrogel

Chitosan/alginate (w/w)	*Crosslinker (wt%) in the original mixture*	*Swelling medium*	*Equilibrium mass swelling*
33/67	1.5	water	66
		PBS	43
	3.1	water	51
		PBS	37
25/75	1.5	water	41
		PBS	39
	3.1	water	36
		PBS	33

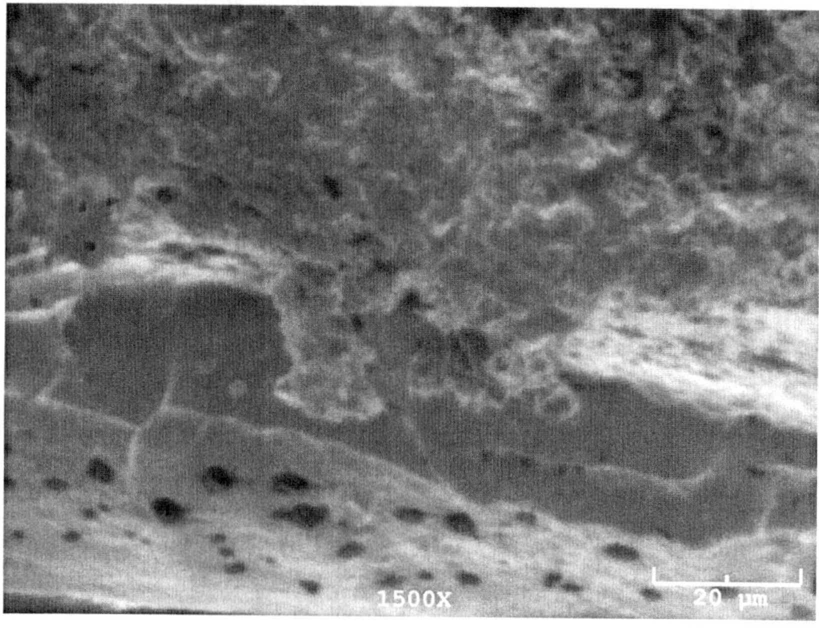

Figure 2. SEM micrograph of the surface and cross section morphology of chitosan/alginate (1/2).

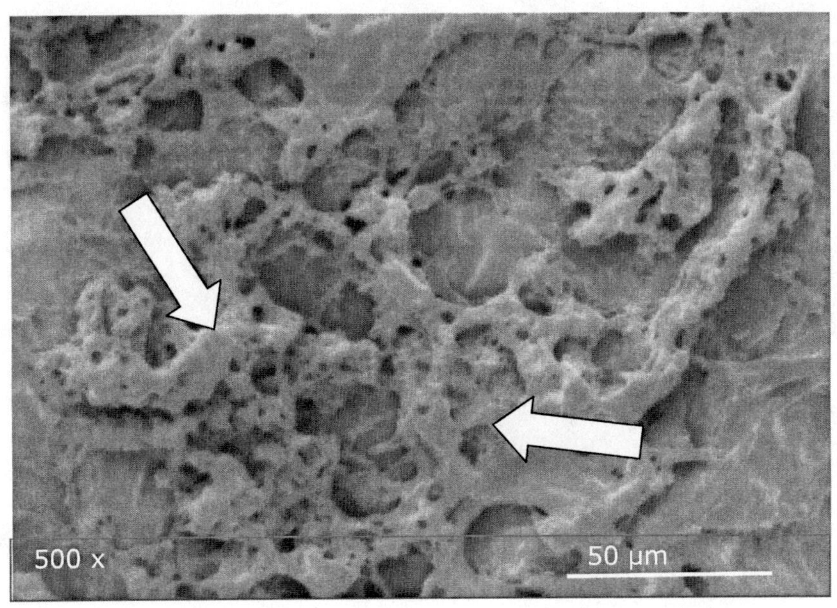

Figure 3. SEM micrograph of the surface morphology of chitosan/alginate (1/2) after being seeded with L929 cells (x500).

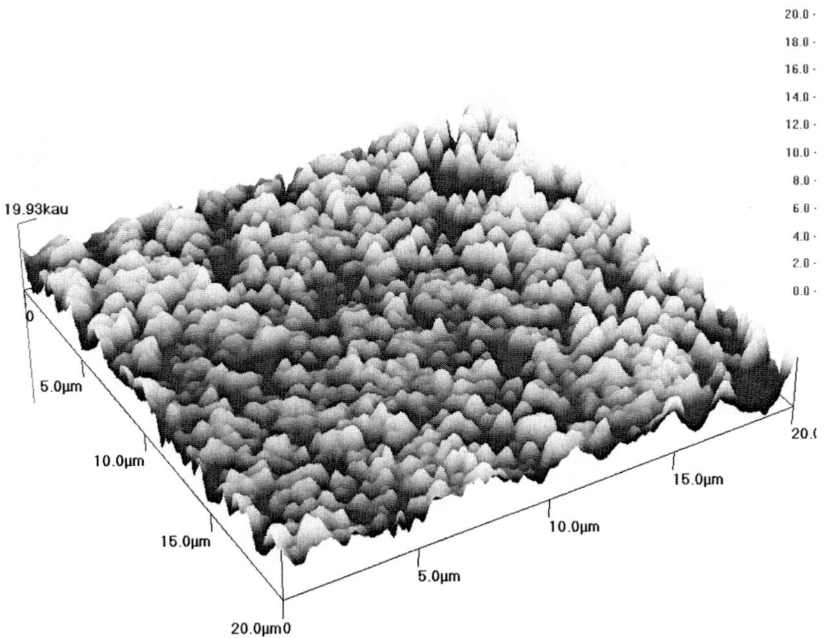

Figure 4. AFM micrograph of chitosan/alginate (1/2) hydrogel.

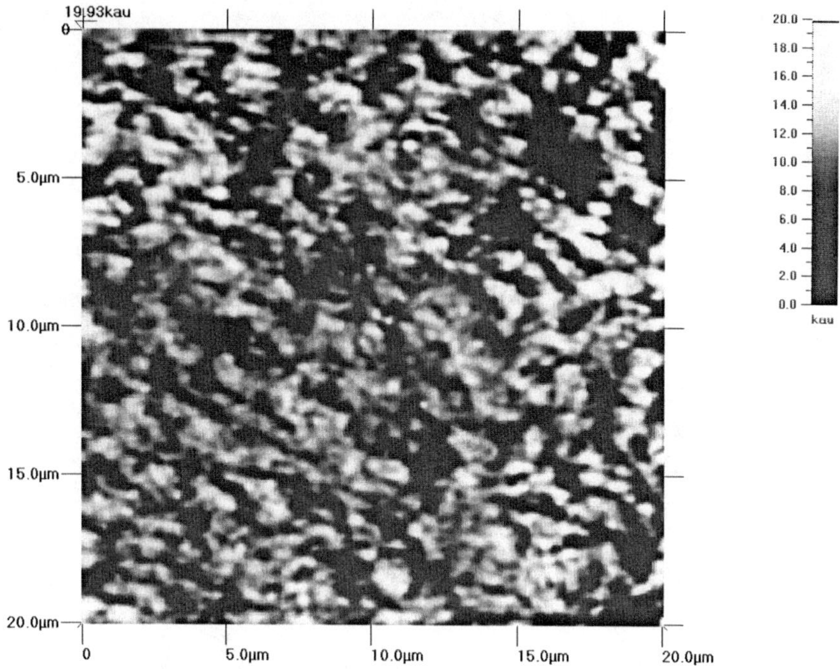

Figure 5. Figure 4. AFM micrograph of chitosan/alginate (1/2) hydrogel.

Conclusions

We studied two hydrogels, chitosan-polycaprolactone and chitosan-alginate.

Lyophilizied hydrogel samples were used in-vitro cell culture experiments with L929 mouse fibroblasts to determine the effects of these polymeric gels on cell attachement and growth.

Cell growth experiments were carried out using chitosan/PCL gels, but it was not succesfull because the crosslinked gels used as scaffolds were not stable. The acrylate containig chitosan/PCL/acrylate gels were used for this purpose.

On the other hand chitosan/alginate gels were used succesfulluy for cell growth purposes. The details of cell-growth experiments will be published elsewhere.

Drug loading and release experiments were carried out in phosphate buffer at pH 7.4 at room temperature using chitosan/PCL and chitosan/PCL/acrylate gels. Details of drug loading and release experiments using Lidocaine as the model drug will also be published elsewhere.

References

1. Ma, L.; Yu, W.; Ma, X. J. *Appl. Polym. Sci.* **2007**, *106*, 394–399.
2. She, H.; Xiao, X.; Liu, R. *J. Matter Sci.* **2007**, *42*, 8113–8119.
3. Grodzinski, J. *J. Polym. Adv. Technol.* **2006**, *17*, 395–418.
4. Vieira, E. F. S; Cestari, A. R.; Airoldi, A. R. *Biomacromolecules* **2008**, *9*, 1195–1199.
5. Cui, Y. L; Qi, A. D; Liu, W. G.; Wang, X. H.; Wang, H.; Ma, D. M. *Biomaterials* **2003**, *24*, 3859.
6. Hsu, S. H.; Whu, S. W.; Tsai, C. L.; Wu, Y. H.; Chen, W. H.; Hsieh, K. H. *J. Polym. Res.* **2004**, *11*, 141.
7. Guibal, E. *Sep. Purif. Technol.* **2004**, *38*, 43–74.
8. Muzzarelli, R A. A. *Carbohydr. Polym.* **2009**, *77* (1), 1–9.
9. Julian, T. N.; Radebaugh, G. W.; Wisniewski, S. J. *J. Controlled Release* **1988**, *7*, 165–169.
10. Li, Z.; Ramay, H. R.; Hauch, K. D.; Xiao, D.; Zhang, M. *Biomaterials* **2005**, *26*, 3919.
11. Li, Z.; Zhang, M. J. *Biomed. Mater. Res.* **2005**, *75A*, 485.
12. Rinaudo, M. Rewiew. *Polym. Int.* **2008**, *57*, 397.
13. Peppas, N. A.; Hilt, J. Z.; Khademhosseini, A.; Langer, R. *Adv. Mater.* **2006**, *18*, 1345.
14. Drury, J. L.; Money, D. J. *Biomaterials* **2003**, *24*, 4337.
15. Rosa, D. S.; Guedes, C. G. F.; Pedroso, A. G. ; Calil, M. R. *Mater. Sci. Eng.* **2004**, *24*, 663–670.
16. Chun, Y. S.; Kyung, Y. J.; Jung, H. C.; Kim, W. N. *Polymer* **2000**, *41*, 8729.
17. Sarasam, A.; Madihally, S. V. *Biomaterials* **2005**, *26*, 5500–5508.
18. Kayaman–Apohan, N.; Karal–Yilmaz, O.; Baysal, K.; Baysal, B. M. *Polym.* **2001**, *42*, 4109–4116.
19. Kayaman−Apohan, N.; Baysal, B. M. *Macromol. Chem. Phys.* **2001**, *202*, 1182–1188.
20. Porjazoska, A.; Karal - Yilmaz, O.; Cvetkowska, M.; Baysal, K.; Baysal, B. M. *J. Biomater. Sci., Polym. Ed.* **2002**, *13*, 1119–1134.
21. Tasdelen, B.; Kayaman−Apohan, N.; Güven, O.; Baysal, B. M. *Int. J. Pharm.* **2004**, *278*, 343.
22. Tasdelen, B.; Kayaman−Apohan, N.; Güven, O.; Baysal, B. M. *Phys. Chem.* **2005**, *73*, 340–345.

Chapter 8

Characterization of Poly(vinyl chloride)/Bentonite Nanocomposite Prepared via Melt Blending Method

Nanocomposites via Melt Blending

Selcan Karakus,[1] Jaroslava Budinski-Simendic,[2] Ljljiana Korugic-Karasz,[3] and Ayse Z. Aroguz[1,*]

[1]Department of Chemistry, Engineering Faculty, University of İstanbul, 34320 Avcılar-Istanbul, Turkey
[2]Faculty of Technology, University of Novi Sad, Serbia
[3]Department of Polymer Science and Engineering, University of Massachusetts, Amherst, MA 01002, USA
[*]aroguz@istanbul.edu.tr

Poly(vinyl chloride) based nanocomposites were prepared by blending with organically modified bentonite particles at three different temperatures 160°C, 170°C, 180°C to compare the temperature effect on the morphology of the composites. The unmodified bentonite was also used as the filler material to prepare poly(vinyl chloride)/unmodified bentonite for comparison. The thermal properties of the samples prepared in this work were examined by differential scanning calorimetery and thermal gravimetric analysis. The results of differential scanning calorimetery showed the existence of the single glass transition temperature for all compositions and in comparison with pristine poly(vinyl chloride) the composites made out to be higher glass transition temperatures. The thermal degradation temperature of the composites also increased up to 60°C in comparison with pristine poly(vinyl chloride). The dispersion and interfacial compatibility of bentonite in poly(vinyl chloride) matrix was characterized by scanning electron microscope.

Introduction

In recent years polymer/clay nanocomposites have gained increasing attention of researchers from the academic and industrial area because of their very good properties. The thermal, optical, mechanical, physicochemical and gas barrier properties when compared with pristine polymer or conventional composites are substantially improved (*1–4*). Nanocomposites are prepared by organic fillers as bentonite, montmorilonite or hectorite clay dispersed in the polymer matrix at the nanoscale level (*5*). There are many publications on nanocomposites preparation (*6–10*). The thermal behavior of poly (ε-caprolactone)/nanocomposites was investigated by Pucciariello et al, (*7*, *8*). It was found that the thermal behavior of these nanocomposites depends on the dispersion of the inorganic filler in the polymeric matrix. Organoclay nanocomposites based on natural rubber were prepared and characterized by Lopez Manchado and co-workers (*6*). Recently, Matthew et al., prepared the polymer nanocomposite scaffolds from bioactive ceramic nanoparticles in a chitosan matrix for the purpose of tissue engineering applications (*10*). Smectite clay is the most commonly used mineral for the preparation of the nanocomposites because of ready dispersibility of its layers in suitable media. Bentonite is a layered smectite clay and it can be used in the nanocomposite materials as the filler. Because of the –OH groups at the clay surface, it needs to be modified with organic surfactant to provide compatibility with polymer matrix (*11*, *12*). Polymer/clay nanocomposites have been investigated by using polymers such as polypropylene (*13*), polystyrene (*14*), polymethylmetacrylate (*14*, *15*), polyimide (*16*), poly (vinyl chloride) (PVC) (*17*, *18*). PVC having low thermal stability is one of the important polymers that is extensively used as the polymer matrix. To improve its stability by adding additives, heat stabilizers, and clay fillers either polymerization or melt blending processes were used (*19*). In the present study PVC/bentonite nanocomposites were prepared by melt blending of PVC with organically modified and unmodified bentonite. The effects of clay content (1% and 5%w) and preparing temperature (160°C, 170°C, 180°C) on the nanocomposites structures were investigated. The thermal properties of nanocomposites were also studied. The results obtained for composites PVC/organically modified clay and PVC/ unmodified clay were compared.

Experimental

Materials

Poly(vinyl chloride) with the medium molecular weight (inherent viscosity 0.92, relative viscosity 2.23) was purchased from Aldrich Chemical company, USA. Bentonite was obtained from MTA, Ankara. Cetyltrimethylammonium bromide (CTAB) was supplied from Merck. Dioctylphatalate (DOP) was obtained from Aldrich Chemical company. All materials were used as received.

Methods

Modification of Bentonite

The chemical composition of bentonite was analyzed by using elemental analyzer according to the methods described in literature (*20*) and the results were shown in Table 1.

The method of the modification of bentonite developed elsewhere (*21*) was applied. To separate impurities, ten grams of bentonite was dispersed in one liter distilled water by vigorous agitation at room temperature. Modifier, 0.2 g CTAB in the form of aqueous solution (20 gL^{-1}) was slowly added into dispersion and another 24 hours agitation was applied. The organically modified bentonite, (OMB), was isolated by filtration and repeatedly washed with distilled water until no Br ions were detected in the wash water by using $AgNO_3$ solution. The filtration product OMB was dried under vacuum at room temperature until constant weight. Then it was ground into a fine powder and sieved through a 200 mesh screen.

Preparation of PVC/Bentonite Nanocomposites

PVC/bentonite nanocomposites were prepared by a melt blending process. A mixture of PVC and OMB (1% and 5%w) was sonicated for one hour before a certain amount of DOP plasticiser was added. The samples were molded into sheets 1 mm in thick by hot-pressing (20 Mpa press) at different temperatures (160°C, 170°C, 180°C), for 20 min followed by cooling to room temperature. The sheets were prepared for morphological analysis. The PVC/OMB samples prepared at 180°C were also used for the measurements of the thermal properties. The same procedure was used for the preparation of PVC/unmodified bentonite (PVC/UMB).

Table 1. The chemical composition of bentonite

Constituents	*Percent by weight*
SiO_2	56.28
$Al_2 O_3$	26.79
Fe_2O_3	4.00
MgO	3.67
CaO	1.34

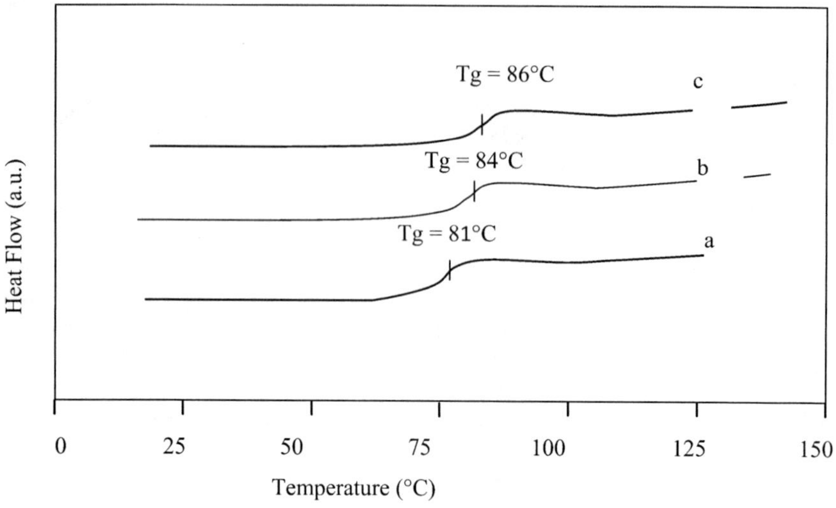

Figure 1. The glass transition temperature of (a) pure PVC (b) PVC/UMB (5wt%) (c) PVC/OMB (5 wt%)

Thermal Analysis

The thermal properties of the nanocomposites were determined by differential scanning calorimetery (DSC Perkin Elmer Model 7) at a heating rate 10 °C/min in temperature range from 25°C to 250°C under nitrogen purge. The glass transition temperature was taken as the initial onset of the change of slope in the DSC curve. The reported Tg values were based on the second run. DSC results are given in Fig 1. The thermo-gravimetric analysis (TGA) were obtained by using a TGA (TGA 2950) instrument. The samples were heated at the heating rate 20 °C/min under a stream of N_2 from 0 °C to 600 °C. The TGA results for the PVC/OMB nanocomposite for different compositions and PVC/UMB nanocomposite are given in Fig. 2 and Fig. 3, respectively.

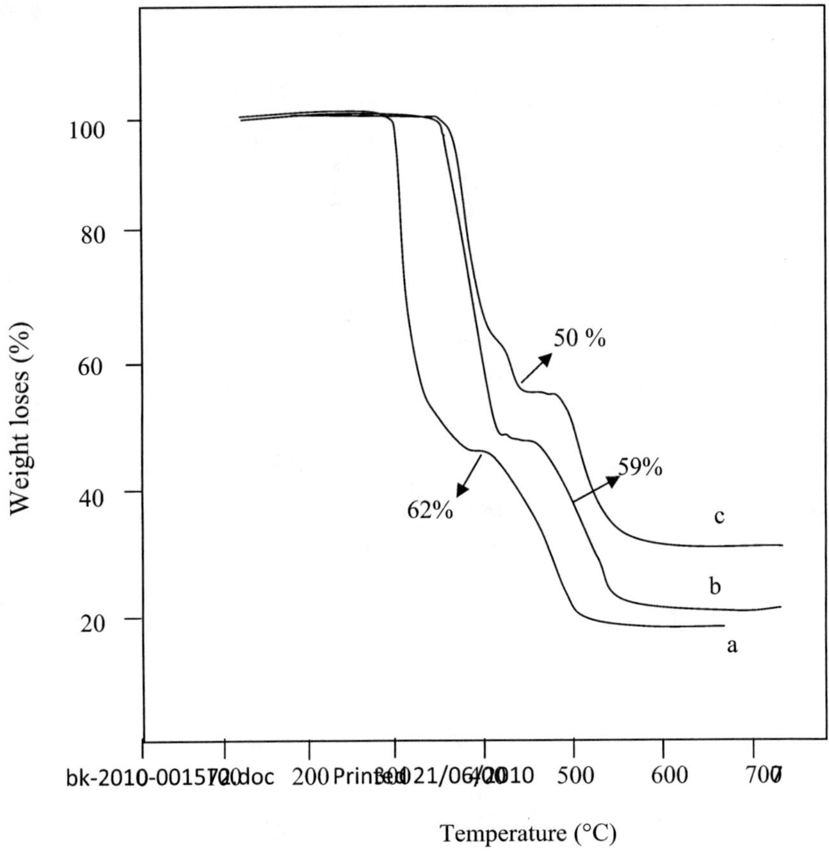

Figure 2. The thermogravimetrical analysis of the samples (a) pure PVC (b) PVC/OMB (1 wt%) (c) PVC/OMB (5 wt%)

Morphological Analysis

The micrographs of the samples were obtained by using scanning electron microscopy (SEM, Jeol JSM6320-Fxv). The specimens were prepared by freeze-fracturing in liquid nitrogen and applying a gold coating of approximately 300 Å on an Edwards S 150B sputter coater. The coated samples were placed in the vacuum chamber of the instrument and observed at the voltage of 5 kV. Fig.4 shows the structure of organically modified bentonite. SEM images of the composites are shown in Figs 5, 6, 7, 8, and 9.

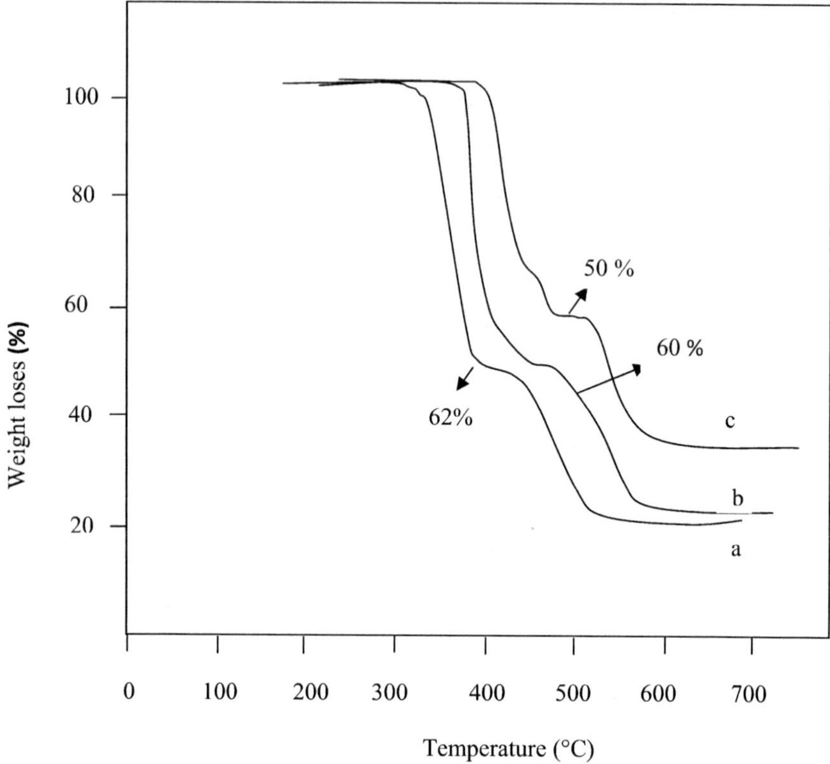

Figure 3. The thermal analysis of the samples (a) pure PVC (b) PVC/UMB (5 wt%) (c) PVC/OMB (5 wt%)

Results and Discussion

Thermal Properties

The glass transition temperatures (Tg) of pure PVC and composites with unmodified and modified filler were at 81 °C, 84 °C and 86 °C, respectively (Fig. 1). Tg of PVC composites with OMB are shifted to the higher temperature when compared with unfilled PVC. The thermal stability of nanocomposites with different clay loadings (1wt %, 5wt %) were investigated and the results are given in Fig 2. Two different thermal degradation steps between 20°C-500°C were observed for pure PVC. PVC was degraded at 298°C and the weight loss of PVC at this step was observed to be 62 wt%. In the first degradation step HCl is generated as reported in the literature (*22*). Two different thermal degradation steps were also obtained for the nanocomposite in the same temperature range. The first decomposition temperature of PVC/OMB nanocomposite (1 wt%) was observed at 354 °C. It is higher than pure PVC by about 56 °C. The first decomposition temperature for nanocomposite (5 wt%) was also observed at 359 °C. The thermal stability of the nanocomposites increased with increasing amount of bentonite in the sample.

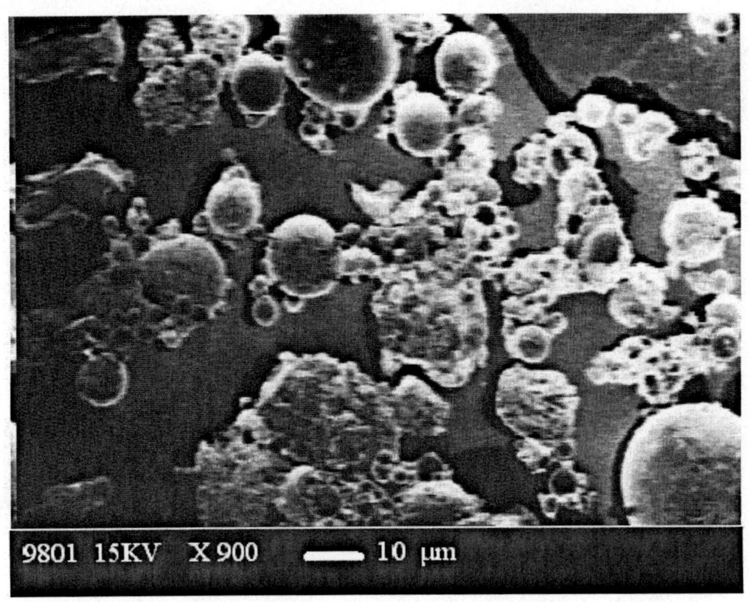

Figure 4. SEM micrographs of organically modified bentonite.

Figure 5. SEM micrographs of the fractured surface for PVC/OMB nanocomposite (1 wt%) prepared at 160 °C

Figure 6. SEM micrographs of the fractured surface for PVC/OMB nanocomposite (5 wt%) prepared at 160 °C.

PVC/OMB nanocomposites showed greater improvement on the thermal properties than the samples prepared with UMB. Almost 50 wt% degradation was observed at the first step of thermo-gravimetric diagram of PVC/OMB (5 wt%). PVC/UMB (5 wt%) nanocomposites in comparison, started to degrade at 354 °C and 60% of the sample was lost in the first step (Fig 3).

Morphological Studies

The morphological analysis and nanophase dispersion were studied by using SEM (Figs. 45678). Fig 4 shows the observation of the organically modified bentonite. SEM micrographs of the composites prepared at various temperatures with two different OMB contents were also shown in Fig 5678Fig 9. Comparing these composites (1 wt% and 5 wt%) it can be seen that the morphology of the fractured surface is looser and rougher in Fig 5. The more homogeneous structure was observed in composites with 5 wt% OMB for all preparation temperatures. The agglomeration also increases with higher OMB content in nanocomposites (Fig 6.) Similar conclusions are confirmed from the comparison of Fig 7 and Fig 8. The more rigid structure was obtained for 5 wt% content of OMB. The morphology of composites changed at high preparation temperatures (Fig 9).

Figure 7. SEM micrographs of the fractured surface for PVC/OMB nanocomposite (1 wt%) prepared at 170 °C.

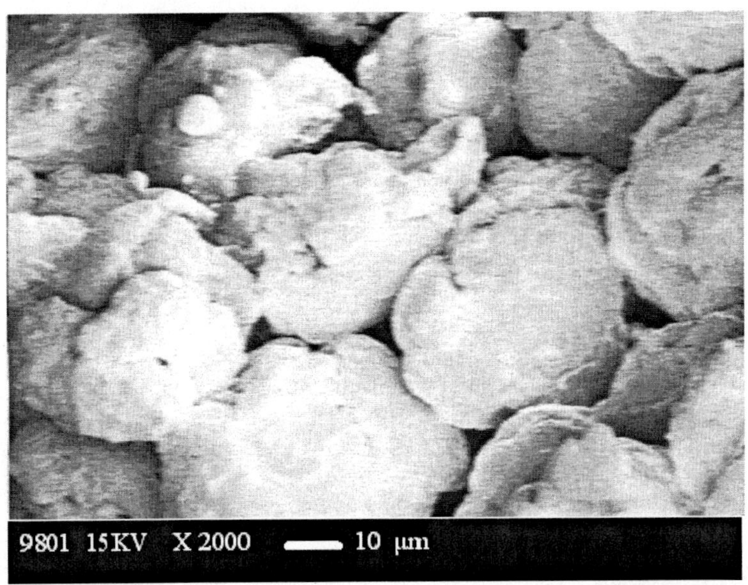

Figure 8. SEM micrographs of the fractured surface of PVC / OMB (5wt%) prepared at 170 °C.

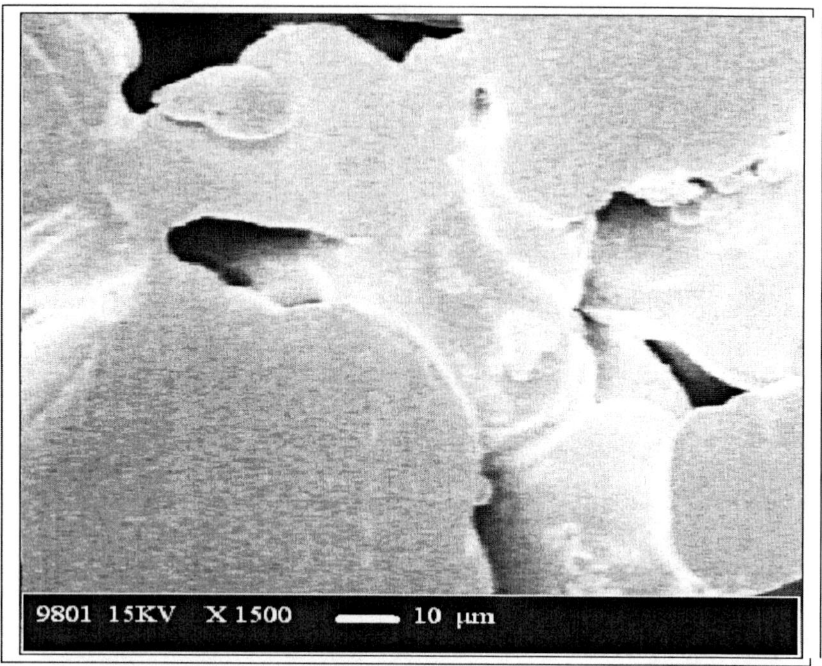

Figure 9. SEM micrographs of the fractured surface of PVC / OMB (5wt%) prepared at 180 °C.

Conclusions

PVC/bentonite nanocomposites were prepared via melt blending method. PVC was blended with organically modified (1 wt% and 5 wt%) and unmodified (5 wt%) bentonite at three different temperatures (160°C, 170°C, 180°C). The effects of clay content and preparing temperature on the nanocomposites structure were compared. The thermal stability of the nanocomposites increased with the higher clay content in the nanocomposites. Organically modified bentonite in nanocomposite improved the thermal stability of PVC more than unmodified bentonite in nanocomposite. High degradation temperature was obtained at higher clay content. The morphology of the samples prepared at different melt blending temperatures was examined and it was found that the structure of composites changed at 180°C. The size of observed particles also increased with higher OMB content in nanocomposites.

References

1. Leszczyńska, A.; Njuguna, J.; Pielichowski, K.; Banerjee, J. R. *Thermochim. Acta.* **2007**, *453* (2), 1, 75–96.
2. Meneghetti, P.; Qutubuddin, S. *Thermochim. Acta.* **2006**, *442* (1-2), 74–77.

3. Pavlidou, S.; Papaspyrides, C. D. *Prog. Polym. Sci.* **2008**, *33* (12), 1119–1198.
4. Song, L.; Qiu, Z. *Polym. Degrad. Stab.* **2009**, *94* (4), 632–637.
5. Moreno, M.; Santa Ana, M. A.; Gonzalez, G.; Benavente, E. *Electrochim. Acta* **2010**, *55* (4), 1323–1327.
6. López-Manchado, M. A.; Herrero, B.; Arroyo, M. *Polym. Int.* **2004**, *53*, 1766–1772.
7. Pucciariello, R.; Villani, V.; Guadagno, L.; Vittoria, V. *J. Polym. Sci., Part B: Polym. Phys.* **2006**, *44*, 22–32.
8. Pucciariello, R.; Villani, V.; Belviso, S.; Gorrasi, G.; Tortora, M.; Vittoria, V. *J. Polym. Sci., Part B: Polym. Phys.* **2004**, *42*, 1321–1332.
9. Stamatialis, D. F.; Papenburg, B. J.; Gironés, M.; Saiful, S.; Bettahalli, S. N. M.; Schmitmeier, S.; Wessling, M. *J. Membr. Sci.* **2008**, *308*, 1–34.
10. Mathew, P.; Binulal, N. S.; Soumya, S.; Nair, S. V.; Furuike, T.; Tamura, H.; Jayakumar, R. *Carbohydr. Polym.* **2010**, *79*, 284–289.
11. Ray, S. S.; Okamoto, M. *Prog. Polym. Sci.* **2003**, *28*, 1539–1641.
12. Alexandre, M.; Dubois, P. *Mater. Sci. Eng., R* **2000**, *28*, 1–63.
13. Shariatpanahi, H.; Sarabi, F.; Mirali, M.; Hemmati, M.; Mahdavi, F. *J. Appl. Polym. Sci.* **2009**, *113* (2), 922–926.
14. Zeng, C.; Lee, L. J. *Macromolecules* **2001**, *34*, 4098–4103.
15. Kumar, S.; Jog, J. P.; Natarajan, U. *J. Appl. Polym. Sci.* **2003**, *89*, 1186–1194.
16. Jang, B. N.; Wilkie, C. A. *Polymer* **2005**, *46*, 3264–3274.
17. Yalcin, B.; Cakmak, M. *Polymer* **2004**, *45*, 6623–6638.
18. Ren, J.; Huang, Y.; Lui, Y.; Tang, X. *Polym. Test.* **2005**, *24*, 316–323.
19. Gang, F.; Feng, M.; Zhao, C.; Zhang, S.; Yang, M. *Polym. Degrad. Stab.* **2004**, *84*, 289–294.
20. Bereket, G.; Aroguz, A. Z.; Ozel, M. Z. *J. Colloid Interface Sci.* **1997**, *187*, 338–343.
21. Chakraborty, S.; Sengupta, R.; Dasgupta, S.; Mukhopadhyay, R.; Bandyopadhyay, S.; Joshi, M.; Ameta, S. C. *J. Appl. Polym. Sci.* **2009**, *113* (2), 1316–1329.
22. Zhu, H. M.; Jiang, X. G.; Yan, J. H.; Chi, Y.; Cen, K. F. *J. Anal. Appl. Pyrolysis* **2008**, *82*, 1–9.

Self-Assembly and Polymer Networks

Chapter 9

Supramolecular Triblock Copolymer Complexes

Gerrit ten Brinke,* Gerrit Gobius du Sart, Ivana Vukovic, Evgeny Polushkin, and Katja Loos

Zernike Institute for Advanced Materials, University of Groningen, Nijenborhgh 4, 9747 AG Groningen, The Netherlands
***brinke@chem.rug.nl**

Four different poly(*tert*-butoxystyrene)-*b*-polystyrene-*b*-poly(4-vinylpyridine) (P*t*BOS-*b*-PS-*b*-P4VP) linear triblock copolymers, with a P4VP weight fraction varying from 0.08 to 0.39, were synthesized *via* sequential anionic polymerization. The values of the unknown interaction parameters between styrene and *tert*-butoxystyrene and between *tert*-butoxystyrene and 4-vinylpyridine were determined from random copolymer blend miscibility studies. All triblock copolymers synthesized adopted a P4VP/PS core/shell cylindrical self-assembled morphology. From these four triblock copolymers supramolecular complexes were prepared by hydrogen bonding a stoichiometric amount of pentadecylphenol (PDP) to the P4VP blocks. Three of these complexes formed a triple lamellar ordered state with additional short length scale ordering inside the P4VP(PDP) layers. The difference in morphology between the triblock copolymers and the supramolecular complexes is due to the change in effective composition and the reduction in interfacial tension between the PS and P4VP containing domains.

Introduction

In the field of nanotechnology, many possible applications are foreseen due to the unique properties of nanoscale objects. An important tool in this field is the self-assembly of block copolymers, which is therefore an intensively researched topic in polymer science nowadays. For simple diblock copolymers, a variety of morphologies has been found, and this number has dramatically increased with the investigations of tri- and multiblock copolymers as well as star-type topologies (*1*). Of course, an important issue is whether these morphologies will actually lead to interesting applications. One promising feature of ABC linear triblock copolymer self-assembly is the possibility to separate A- and C- phases with a B phase, where the latter may be in the form of either the matrix (or shell) for cylindrical, spherical or bincontinuous morphologies or as lamellae in the case of triple-lamellar systems. Our research was partially motivated with the possibility to prepare nanoporous membranes in mind. To investigate new systems in which two phases are separated from each other by a third phase, we synthesized novel linear ABC triblock copolymers, notably poly(4-*tert*-butoxystyrene)-*b*-poly(styrene)-*b*-poly(4-vinylpyridine) or P*t*BOS-*b*-PS-*b*-P4VP. The triblock copolymers are synthesized *via* a three-step sequential anionic polymerization (Figure 1).

In these triblock copolymers, all phases have interesting properties, which may prove useful for many different applications. Firstly, as was investigated thoroughly in the past (*2–4*), the poly(4-vinylpyridine) (P4VP) phase allows the complexation of low molecular weight amphiphiles such as pentadecylphenol (PDP) through hydrogen bonding, which in the case of block copolymers results in so-called hierarchical structure formation . This is due to the fact that within the hydrogen bonded side-chain P4VP(PDP) domains further microphase separation occurs between the alkyl tails of the amphiphiles and the rest of the comb block. Of course, the addition of these amphiphiles will dramatically change both the interactions associated with the P4VP phase as well as increase its weight fraction. Therefore, it is of direct interest to investigate the differences in self-assembly between the bulk triblock copolymer and its supramolecular complexes with PDP (Figure 2). In relation to membrane applications (*5*), the most interesting property is the possibility to wash away these amphiphiles after the structure formation has taken place, which will then create a nanoporous structure. This phase may subsequently be, e.g., positively charged by quarternization at the nitrogen atom, which could be of interest for ion separation from fluids.

It was shown before that the *tert*-butyl group of P*t*BOS can be removed in solution, thus creating a poly(hydroxystyrene) macromolecule (*6*). Although it is known that this removal is more difficult than the removal of a *tert*-butyl methacrylate group, it may be possible to remove the *tert*-butyl group in thin films. This would then give another porous phase. If this is achieved, a membrane with two different types of pores and pore sizes may be created from the original triblock copolymers. The PS phase is not easily derivatized by aforementioned types of reactions, let alone degraded. Therefore, it will serve as a stable separating phase for the other block domains.

To rationalize the observed self-assembly of these systems, additional investigations into the values of the pertinent interaction parameters were conducted. From previous miscibility studies on blends of PS with P(S-*co*-4VP) random copolymers (*7*) it was concluded that $0.30 < \chi_{S,4VP} \leq 0.35$. The other two interaction parameters, $\chi_{S,tBOS}$ and $\chi_{4VP,tBOS}$ are determined by similar blend

sBuLi $\xrightarrow{tBOS}$ $\xrightarrow{St}$ $\xrightarrow{1)\ 4VP\quad 2)\ CH_3OH}$

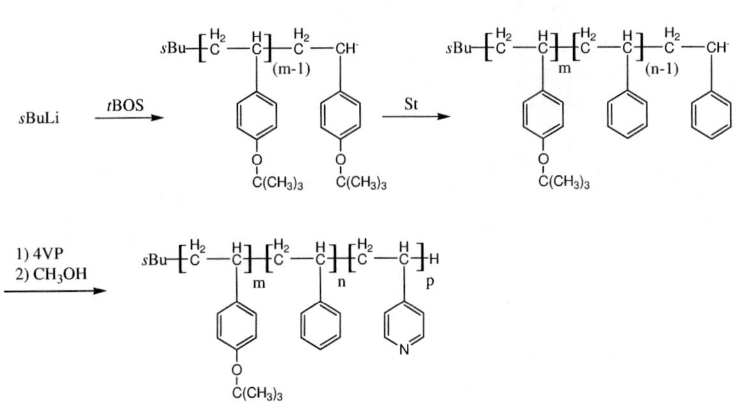

Figure 1. Synthesis route of the PtBOS-b-PS-b-P4VP triblock copolymer.

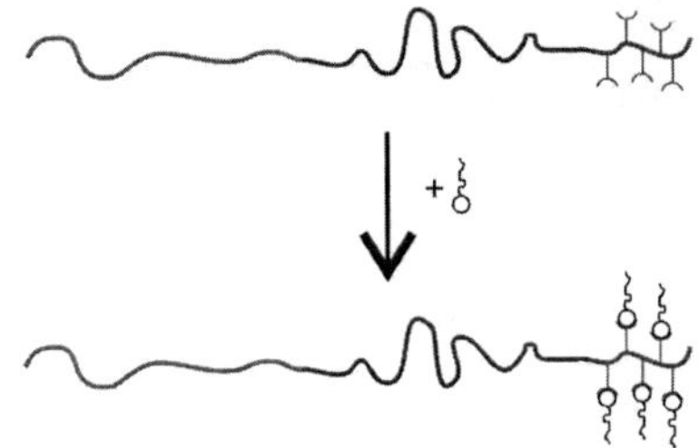

Figure 2. Triblock copolymer supramolecular complexes.

studies presented here and, as will be shown, satisfy $0.031 < \chi_{S,tBOS} < 0.034$ and $0.39 < \chi_{4VP,tBOS} < 0.43$. This implies that the self-assembly in these triblock copolymers is expected not to show any frustrated morphologies. Moreover, since $\chi_{AB} << \chi_{BC}$, core-shell self-assembled states are to be expected.

The triblock copolymer morphology is greatly affected by adding PDP. For the pure triblock copolymer $\chi_{S,tBOS} << \chi_{S,4VP}$ holds. After addition of PDP the interaction involves PS and P4VP(PDP). From previous studies (*8*) it is known that the presence of the PDP side chains reduces the interaction considerably. Together with the fact that the weight fraction of the P4VP phase increases significantly on addition of PDP, its presence will obviously strongly affect the final morphology. In particular self-assembled states with an increased interface between PS and P4VP(PDP), compared to PS and P4VP, are to be expected.

Interactions Involved

Over the years, different researchers have used different methods to determine χ-parameters experimentally. Among these methods are: neutron reflectivity experiments (*9*), small-angle X-ray scattering regarding the ODT of diblock copolymers (*10*), the solubility parameter approach (*11*) (which is merely a calculation based on known or estimated solubility parameter values) and contact angle measurements (*12*). In the past, we have performed so-called random copolymer miscibility studies to calculate the corresponding χ-parameter (*7, 13, 14*).

According to a simple mean-field analysis the Gibbs free energy of mixing per segment for a blend of homopolymer P(A) and random copolymer P(A_x-*co*-B_{1-x}) is given by (*13*)

$$\frac{\Delta g_m}{kT} = \frac{\varphi}{N_1}\ln\varphi + \frac{1-\varphi}{N_2}\ln(1-\varphi) + \chi_{eff}\,\varphi(1-\varphi) \qquad (1)$$

if both components are assumed to be monodisperse with chain lengths N_1 and N_2, respectively; φ denotes the volume fraction of P(A). The essential observation is that the interaction parameter χ_{eff} is related to χ_{AB} by

$$\chi_{eff} = (1-x)^2\,\chi_{AB} \qquad (2)$$

and thus will be much smaller than χ_{AB} for $x \to 1$. Phase separation occurs for

$$\chi > \frac{1}{2}\left(\frac{1}{\sqrt{N_1}} + \frac{1}{\sqrt{N_2}}\right)^2 \qquad (3)$$

Equation 2 demonstrates that both species will become miscible for x sufficiently close to 1. In general the polymer species involved are not monodisperse. Therefore, the first two terms in eq. 1, representing the translational entropy of the two monodisperse polymers involved, will have to be replaced by many similar terms, one for each chain length. The effect of polydispersity on the miscibility of different polymers has been discussed by Koningsveld *et al.* (*15*) who showed that in a good approximation eq. 3 remains valid provided weight average chain lengths are used.

From these equations it is evident that the value of χ_{AB} can be estimated by investigating the value of x for which immiscibility turns into miscibility for blends of P(A_x-co-B_{1-x}) random copolymers with homopolymer P(A). One of the simplest ways to judge miscibility is by Differential Scanning Calorimetry (DSC). The blend is heated into the melt state (above the glass transitions of both polymers), annealed and cooled to room temperature again. In a subsequent scan miscibility is judged on the basis of the presence of one or two glass transition temperatures. The presence of two glass transition temperatures is a clear indication of macrophase separation having occurred in the melt state. The presence of a single glass transition implies a molecularly mixed state provided the glass transition temperatures of the respective pure components are sufficiently far apart (e.g. 20 °C). Because of the nature of the blends, the glass transition temperatures may, however, often be quite close implying that the above criterion can no longer be used. This is clearly always the case for $x \to 1$. In such cases an alternative procedure, based on physical aging or thermal annealing of the blend system in the glassy state, is available (*16–18*). When a molecularly homogeneous blend of two polymers with comparable glass transition temperatures is annealed at a temperature T_a not too far below their T_gs, a subsequent DSC scan will reveal a single enthalpy relaxation peak. However, when the blend is phase separated, the thermal annealing procedure, if correctly applied, will reveal two enthalpy recovery peaks. This is due to the fact that thermal aging is an extremely complicated process involving a spectra of relaxation times. Therefore, even if two chemically different polymers have a nearly identical glass transition temperature, the effect of physical aging will generally be different and different polymers will exhibit different relaxation peaks. Using modulated DSC, which can separate reversible and non-reversible heat flow signals, the analysis of these physical aging experiments have become unambiguous. The physical aging temperature T_a should not be too far below the T_gs, since this will make the relaxation exceedingly slow. On the other hand, when T_a is too close to the T_gs, the polymer system relaxes quickly to its equilibrium state and only a single shallow enthalpy recovery peak will be visible irrespective of the phase state of the mixture. Practical experience showed that the optimum difference between T_g and T_a varies somewhat, with 15-20 K usually being an appropriate choice.

Since the χ-parameter for interactions between St and 4VP was already determined, we performed such blend studies for the two remaining monomer pairs, St/*t*BOS and *t*BOS/4VP. To this end, random copolymers were synthesized *via* free radical polymerization. Firstly, the results of the PS / P(S_x-*co*-*t*BOS_{1-x}) blend study will be discussed. The characteristics of the synthesized St / *t*BOS random copolymers are listed in Table 1. All random copolymers of St and tBOS showed a Tg around 105 °C and they were blended (50/50 wt%) with a PS sample with a Tg of 92 °C (M_w=7025 g/mol). The DSC scans of these blends are shown in Figure 3.

The difference in Tg is so small, that the nature (single or double) of glass transitions cannot be resolved. Especially the curves for the 18 % and 25 % St in the copolymer show exceptionally broad glass transitions, suggesting the presence of two superimposed Tgs. Therefore, the blends were physically aged at 90 °C for at least 4 days to clarify whether the blend was mixed (one enthalpy recovery peak) or phase separated (two enthalpy recovery peaks). In Figure 4 the subsequent DSC recovery scans of a number of blends are shown.

From the aged curves, we may conclude that copolymers with fractions St of 33 % and above are miscible (one enthalpy recovery peak) with the polystyrene sample of M_w=7025 g/mol. Furthermore, phase separation occurs at St fractions of 30 % and below (two enthalpy recovery peaks). From equations (2) and (4) it now straightforwardly follows that:

$$0.031 < \chi_{S,tBOS} < 0.034 \quad (4)$$

A similar blend study was performed to determine the interaction parameter between *t*BOS and 4VP. The copolymers synthesized are listed in Table 2. The random copolymers were blended (50/50 wt%) with a poly(*tert*-butoxystyrene) homopolymer with M_w=71.9 kg/mol. The reversible heat flow curves of the blends are shown in Figure 5.

Table 1. Characteristics of the St / *t*BOS random copolymers

Entry	*Fraction St* [a] *(weight%)*	M_w [b] *(kg/mol)*	*PDI* [b]	*Tg (°C)*
P*t*BOS	0	59.2	1.53	108
1	17.9	35.0	1.44	107
2	24.9	37.0	1.46	107
3	30.0	35.8	1.44	108
4	32.6	37.0	1.47	107
5	41.0	34.7	1.46	107
6	49.1	47.3	1.52	109
7	58.0	31.0	1.45	106
8	66.4	31.0	1.44	105
9	72.9	34.0	1.45	104

[a] Determined from 1H-NMR. [b] Determined by GPC in DMF, calibrated against PS.

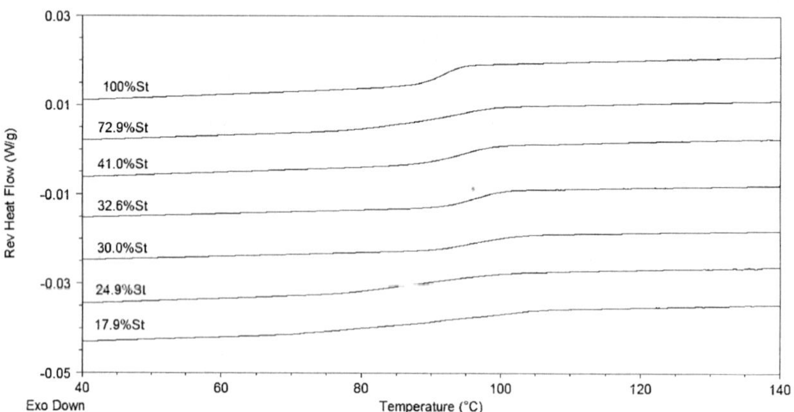

Figure 3. Reversible heat flow curves for 50/50 wt% blends of PS (M_w = 7025 g/mol) with different P(S-co-tBOS) random copolymers.

It is clear from these curves that macrophase separation occurs in blends of the random copolymers with a fraction *t*BOS equal to and below 63 %. To judge the miscibility of the four blends with the highest fractions of *t*BOS, they were aged at 95 °C. The subsequent DSC recovery curves are shown in Figure 6 Two enthalpy recovery peaks are observed for tBOS fractions in the copolymer of 87 %, 77 % and 91 %, indicating a phase separated system. The blends with tBOS fractions in the copolymer of 96 % shows one recovery peak. However, these findings only result in an unacceptably broad range of values from 0.39 to 1.84. To reduce this large window, PtBOS homopolymers were synthesized with different molecular weights. These were subsequently blended with the random copolymer with 91

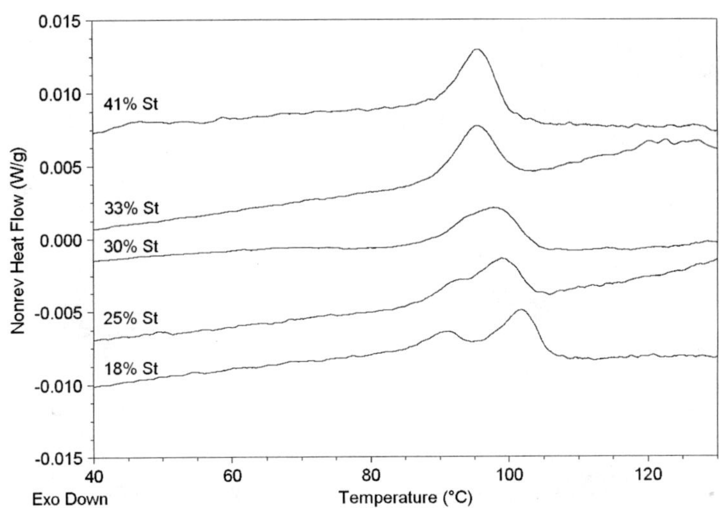

Figure 4. Non-reversible heat flow curves for 50/50 wt% different PS / $P(S_x$-co-$tBOS_{1-x})$ blends after physical aging at 90 °C.

Table 2. Characteristics of the *t*BOS / 4VP random copolymers

Entry	*Fraction tBOS (weight%)* [a]	*M_w (kg/mol)*	*PDI* [b]	*Tg (°C)*
P4VP	0	50.0	NA	152
1	17.2	114.0 [b]	1.45	143
2	34.6	100.1 [b]	1.54	139
3	56.8	85.1 [b]	1.46	134
4	62.6	46.6 [b]	1.45	123
5	77.1	70.0 [b]	1.44	120
6	86.7	61.1 [b]	1.43	111
7	91.1	61.7 [c]	1.41	113
8	96.2	69.7 [c]	1.40	107
PtBOS	100	71.9 [b]	1.55	108

[a] Determined from ^{1}H-NMR. [b] Determined by GPC in DMF, calibrated against PS. [c] Determined by GPC in THF, calibrated against PS.

% tBOS and physically aged at 95 °C for three weeks to determine miscibility (Figure 7).

The 91% *t*BOS random copolymer shows miscibility (one symmetrical peak) with the homopolymer with M_w = 58.9 kg/mol as well as M_w = 24.2 kg/mol. A peak with a shoulder indicating phase separation is observed for the homopolymer with M_w = 71.9 kg/mol (same curve as in Figure 6). Hence, it follows that

$$0.39 < \chi_{4VP,tBOS} < 0.43 \qquad (7)$$

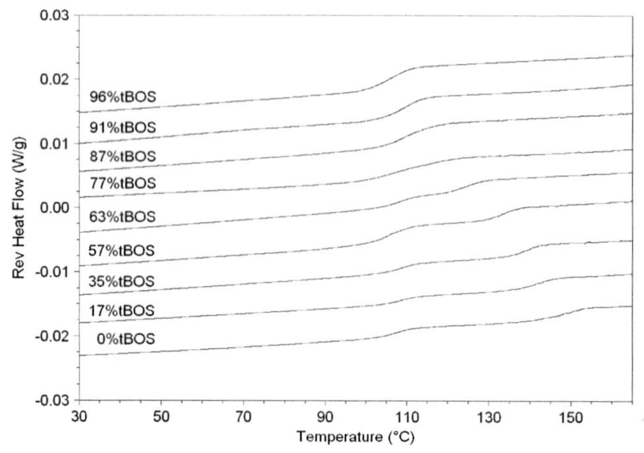

Figure 5. Reversible heat flow curves for different 50/50 wt% PtBOS / P(tBOS$_x$-co-4VP$_{1-x}$) blends.

which is sufficiently accurate for our purpose.

Summarizing we see that the interaction between 4VP and both S and *t*BOS is of the same order and highly unfavorable, whereas the interaction between S and *t*BOS is only slightly unfavorable. Hence, the system will try to minimize the PS/P4VP interface without creating any P4VP/P*t*BOS contacts and we should expect to find core-shell morphologies.

Self-Assembly of P*t*BOS-*b*-PS-*b*-P4VP Triblock Copolymers

The PtBOS-b-PS-b-P4VP linear triblock copolymers were synthesized through a three-step sequential anionic polymerization. For each polymerization step well-defined polymers with a low polydispersity index (PDI) and predictable molecular weights were obtained. Table 3 lists the properties of the four different triblock copolymers synthesized.

Figure 8 shows a TEM picture and the SAXS scattering curve of the self-assembled structure of entry 2 of Table 3. As for the other three triblock copolymer systems a P4VP-PS core-shell cylindrical structure is found thus demonstrating that this structure is stable for a wide range of P4VP fractions. This kind of behavior is supported by a study of Tang *et al.* (*19*) who performed a real space implementation of the self-consistent field theory in a two-dimensional space. They found that for ABC triblock copolymers with $\chi_{AB}N << \chi_{BC}N$ and $\chi_{BC}N$ slightly smaller than $\chi_{AC}N$ (which corresponds perfectly to our case), the core-shell cylindrical morphology is found for a large part of the ternary phase diagram.

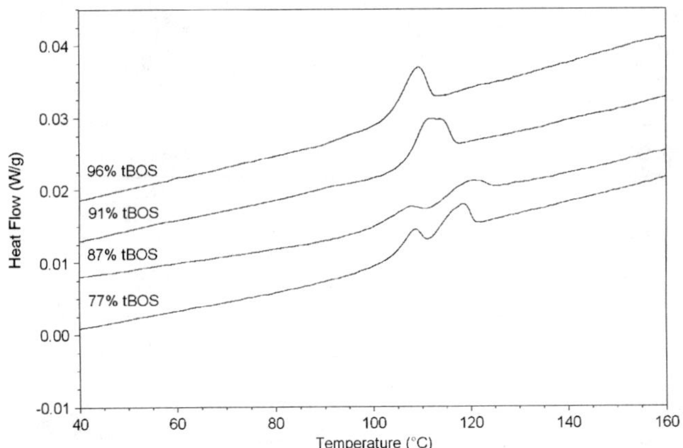

Figure 6. DSC scans of different 50/50 wt% P4VP / P(4VP$_x$-co-tBOS$_{1-x}$) blends after physical aging.

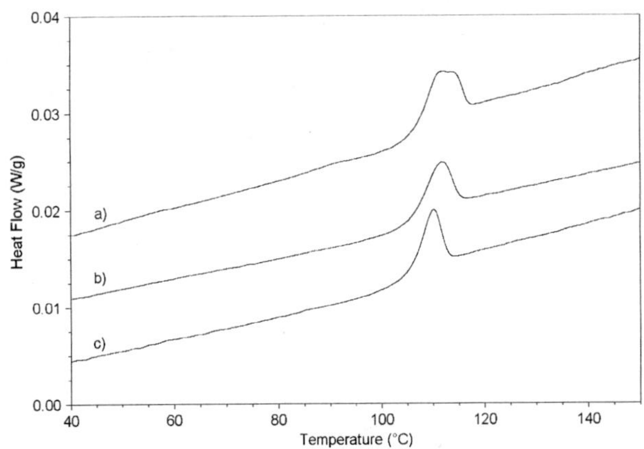

Figure 7. Heat flow curves of P(tBOS$_{0.91}$-co-4VP$_{0.09}$) blended (50/50 wt%) with different PtBOS homopolymers after physical aging at 95 °C. M_w's of the homopolymers were 71.9 kg/mol (a), 59.9 kg/mol (b) and 24.2 kg/mol (c).

The addition of the molecular weight amphiphile pentadecylphenol (PDP) to the triblock copolymers (see Figure 2) changes the triblock copolymer self-assembly in such a way that morphologies with a larger PS / P4VP interface are realized than in the pure block copolymer sample. Characteristically, the triblock copolymers with a core-shell hexagonal cylindrical structure assume a triple lamellar morphology when adding a nominal amount of PDP. These changes are ascribed to the lowering of the PS / P4VP interface tension due to the presence of PDP in the P4VP domains, the increased weight fraction of the P4VP-containing phases and the comb-like nature of the P4VP(PDP) blocks. Figure 9 illustrates this for the P*t*BOS-*b*-PS-*b*-P4VP(PDP) supramolecules based on entry 2 of Table 3.

Table 3. Triblock copolymer properties

	Weight fractions				
Entry	*PtBOS*	*PS*	*P4VP*	M_n *(kg/mol)*	*PDI*
1	0.46	0.46	0.08	76.0	1.04
2	0.29	0.56	0.15	49.7	1.07
3	0.34	0.34	0.32	13.3	1.06
4	0.24	0.38	0.39	90.6	1.09

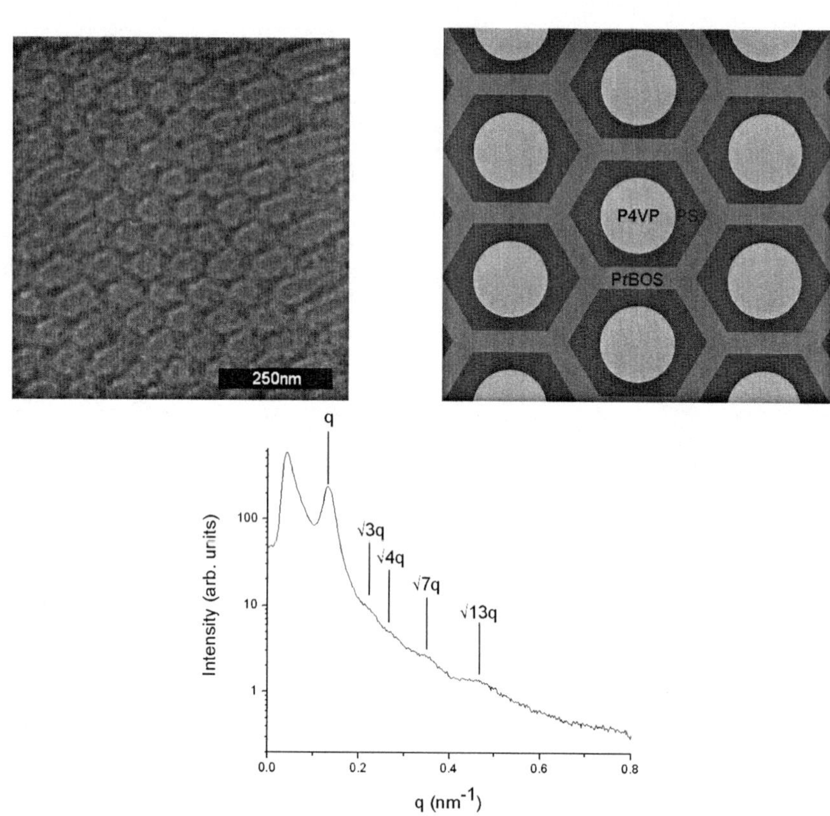

Figure 8. TEM, cartoon of self-assembled structure and SAXS of triblock copolymer 2 (Table 3)

The cartoon below (Figure 10) illustrates the characteristic change in morphology observed on adding PDP to the triblock copolymers synthesized. The short length scale ordering within the P4VP(PDP) domains is also indicated. It is not visible in the TEM pictures but clearly observable in the temperature dependent SAXS measurements at higher angles than presented in Figures 8 and 9 and it is , of course, well-known from our previous diblock copolymer-based supramolecules studies (*20*).

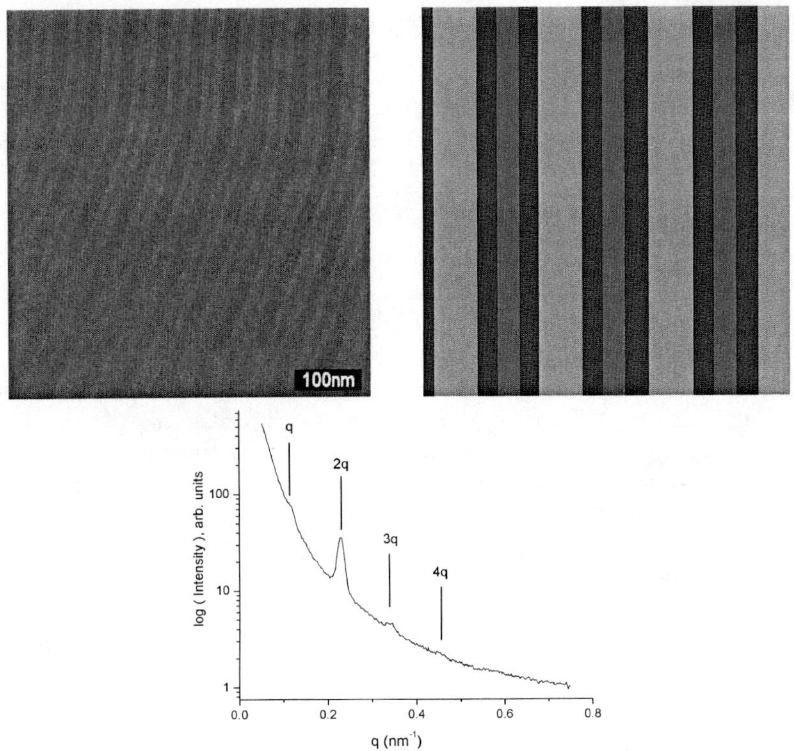

Figure 9. TEM, cartoon of self-assembled structure and SAXS of PtBOS-b-PS-b-P4VP(PDP) supramolecules based on triblock copolymer 2 (Table 3)

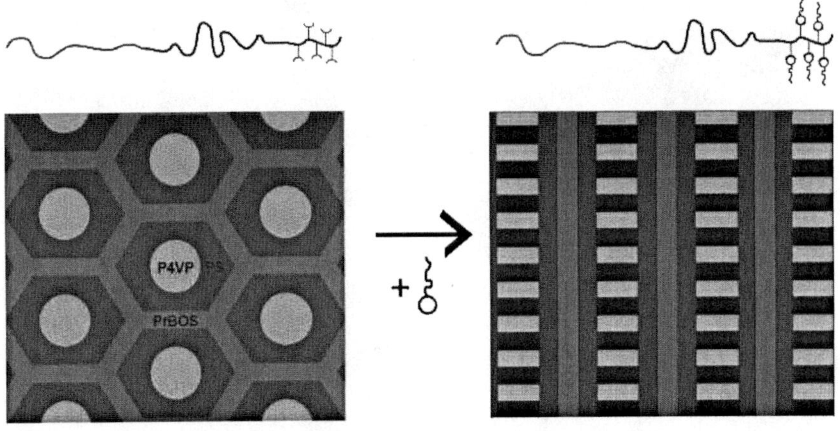

Figure 10. Cartoon of the characteristic morphology of the PtBOS-b-PS-b-P4VP triblock copolymer synthesized (left) and the PtBOS-b-PS-b-P4VP(PDP) supramolecules based thereon.

Concluding Remarks

Four different P*t*BOS-*b*-PS-*b*-P4VP linear triblock copolymers were synthesized and their self-assembly was studied with SAXS and TEM. The P4VP weight fraction varied from 0.08 to 0.39. A random copolymer study was used to determine the values of the two unknown interaction parameters as $0.031 < \chi_{S,tBOS} < 0.034$ and $0.39 < \chi_{4VP,tBOS} < 0.43$, the latter being slightly larger than the $0.30 < \chi_{S,4VP} < 0.35$ already known from a previous study. Given these values, it was not a big surprise to find that all four systems self-assembled in the form of a P4VP/PS core/shell cylindrical morphology. Subsequently the hydrogen-bonded supramolecular complexes of P*t*BOS-*b*-PS-*b*-P4VP with stoichiometric amounts of the low molecular weight amphiphile PDP, i.e. one PDP molecule per 4VP unit, were prepared and investigated. Three of the four showed a triple lamellar morphology, with additional short length scale ordering inside the P4VP(PDP) layers. The ordered state of the sample based on the triblock copolymer with the highest fraction of P4VP, not discussed here (see ref. (*21*)) consisted of alternating layers of P*t*BOS and P4VP(PDP) with PS cylinders inside the latter layer. To obtain self-assembled states with P4VP(PDP) continuous networks, the most interesting state from the perspective of nanoporous membrane preparation, smaller amounts of PDP are required as will be discussed in a future paper (*22*).

References

1. Hamley, I. W. *The Physics of Block Copolymers*; Oxford Science Publications: Oxford, New York, 1998.
2. Ikkala, O.; ten Brinke, G. *Science* **2002**, *295*, 2407.
3. Ikkala, O.; ten Brinke, G. *Chem. Commun.* **2004**, 2131.
4. ten Brinke, G.; Ruokolainen, J.; Ikkala, O. *Adv. Polym. Sci.* **2007**, *207*, 113.
5. Hillmyer, M. A. *Adv. Polym. Sci.* **2005**, *190*, 137.
6. Kuo, S.-W.; Tung, P.-H.; Chang, F.-C. *Macromolecules* **2006**, *39*, 9388.
7. Alberda van Ekenstein, G. O. R.; Meyboom, R.; ten Brinke, G.; Ikkala, O. *Macromolecules* **2000**, *33*, 3752.
8. Polushkin, E.; Alberda van Ekenstein, G. O. R.; Knaapila, M.; Ruokolainen, J.; Torkkeli, M.; Serimaa, R.; Bras, W.; Dolbnya, I.; Ikkala, O.; ten Brinke, G. *Macromolecules* **2001**, *34*, 4917.
9. Schubert, D. W.; Stamm, M.; Müller, A. H. E. *Polym. Eng. Sci.* **1999**, *39*, 1501.
10. Ludwigs, S.; Böker, A.; Abetz, V.; Müller, A. H. E.; Krausch, G. *Polymer* **2003**, *44*, 6815.
11. Zhao, Y.; Sivaniah, E.; Hashimoto, T. *Macromolecules* **2008**, *41*, 9948.
12. Clarke, C. J.; Eisenberg, A.; La Scala, J.; Rafailovich, M. H.; Sokolov, J.; Li, Z.; Qu, S.; Nguyen, D.; Schwartz, S. A.; Strzhemechny, Y.; Sauer, B. B. *Macromolecules* **1997**, *30*, 4184.
13. ten Brinke, G.; Karasz, F. E.; MacKnight, W. J. *Macromolecules* **1983**, *16*, 1824.

14. ten Brinke, G.; Rubinstein, E.; Karasz, F. E.; MacKnight, W. J.; Vukovic, R. *J. Appl. Phys.* **1984**, *56*, 2440.
15. Koningsveld, R.; Chermin, H. A. G.; Gordon, M. *Proc. R. Soc.* London 1970, A319, 331.
16. Bosma, M.; ten Brinke, G.; Ellis, T. S. *Macromolecules* **1988**, *21*, 1465.
17. Grooten, R.; ten Brinke, G. *Macromolecules* **1989**, *22*, 1761.
18. ten Brinke, G.; Grooten, R. *Colloid Polym. Sci.* **1989**, *267*, 992.
19. Tang, P.; Qiu, F.; Zhang, H.; Yang, Y. *Phys. Rev. E* **2004**, *69*, 031803.
20. Ruokolainen, J.; ten Brinke, G.; Ikkala, O. *Adv. Mater.* **1999**, *11*, 777.
21. Gobius du Sart, G.; Vukovic, I.; Alberda van Ekenstein, G. O. R.; Polushkin, E.; Loos, K.; ten Brinke, G. *Macromolecules*, accepted for publication.
22. Gobius du Sart, G.; Vukovic, I.; Loos, K.; ten Brinke, G.; Hiekkataipale, P.; Ruokolainen, J., in preparation.

Chapter 10

Structure and Sorption Properties of Syndiotactic Polystyrene Aerogels

Christophe Daniel* and Gaetano Guerra

Dipartimento di Chimica, Università degli Studi di Salerno, via Ponte Don Melillo, 84084 Fisciano (SA), Italy
***cdaniel@unisa.it**

High porosity monolithic aerogels can be obtained from syndiotatic polystyrene thermoreversible gels by supercritical CO_2 extraction of the solvent present in the native gel. Depending on the gel and on the extraction prodecure, aerogels with different crystalline forms, morphologies and porosities can be obtained. The fast kinetics and high sorption capacity of VOCs from aqueous solutions by sPS aerogels with microporous crystalline phases as well as their good handling characteristics make these new materials particularly suitable as a sorption medium to remove traces of pollutants from water and air. Different aspects relative to the preparation, the structure and the sorption properties of these new polymeric materials are described.

Introduction

Syndiotactic polystyrene (sPS) displays a complex polymorphic behaviour and in the crystalline state five crystalline forms (designed by the acronyms α (*1*, *2*), β (*3*, *4*), γ (*5*, *6*), δ (*7*), and ε (*8*, *9*)) and numerous co-crystal phases (clathrates (*10–15*) and intercalates (*16–18*)) can be obtained with low-molecular-mass compounds.

Of particular interest are the two δ and ε crystalline forms which are characterized by the presence of microcavities in the crystalline unit cell.

The δ-form presents two identical cavities (with an extension of about 0.8 nm and a volume of nearly 120 $Å^3$ (*19*)) per unit cell and is capable to absorb rapidly and selectively volatile organic compounds (VOCs) (mainly halogenated or aromatic hydrocarbons) from water and air, also when present at very low concentrations (*20–23*). Moreover, as these materials are hydrophobic they seem particularly suitable for applications in chemical separation, in air/water purification as well as in sensorics (*24–27*).

The microporous ε crystalline form of sPS is characterized by channel-shaped cavities crossing the unit cells along the *c* axis (*9*), rather than by isolated cavities as observed for the δ form (*7, 19*). The different type of micropores allows the sorption in the ε-form of guest molecules being too large for the cavities of the δ-form (*8*).

SPS presents also the particularity to form thermoreversible gels in a large variety of solvents and depending on the solvent-type and/or thermal treatments two types of gels can be obtained (*28–43*). Elastic gels for which the polymer-rich phase is a co-crystal (clathrate or intercalate) phase are obtained with low-molecular-mass guest molecules (*28–36, 38, 39, 41*) (also called type I gels (*34*)) while with bulky solvent molecules unable to form sPS co-crystals, paste-like solids with much lower elasticity for which the crystalline phase is the thermodynamically stable and dense β-form can be obtained (*34, 37, 40*) (also called type II gels (*34*)).

It has been shown that high porosity aerogels can be obtained by supercritical CO_2 extraction of the solvent present in the native sPS gels (*44–46*) or by sublimation of the solvent (*47, 48*). The advantage of using a supercritical extraction process to remove the solvent from the gel is the absence of surface tension and in these conditions a supercritical solution is formed between supercritical CO_2 and liquid solvent (*49*). Thus, it is possible to extract the solvent from the gel without collapsing the structure and, as an example, Figure 1 shows that the dimensions of a sPS gel remain unchanged during the supercritical CO_2 extraction and total removal of the solvent initially present in the gel.

SPS can form thermoreversible gels for a large range of polymer concentrations and gels with concentrations as low as 0.02 g/g or higher than 0.5 g/g can be obtained. Thus, after solvent extraction with supercritical CO2, monolithic aerogels with apparent density in the range 0.004-0.5 g/cm3 can be easily prepared (see Figure 2).

The fact that the apparent density is much lower than the density of the polymer matrix (c.a. 1 g/cm^3 for polystyrene) indicates that the aerogel is full of open space, or pores.

Qualitatively, the porosity P of aerogel samples can be expressed as a function of the aerogel apparent density ρ as (*50*):

$$P = 100\left(1 - \frac{\rho}{\rho_S}\right)$$

where ρ_S is the density of the polymer matrix (equal to 1.02 g/cm^3, for a semicrystalline δ-form sPS with a crystallinity of nearly 40%)

Thus, for gels prepared at very low polymer concentration, monolithic aerogels with porosity as high as 99.5% can be obtained after solvent extraction.

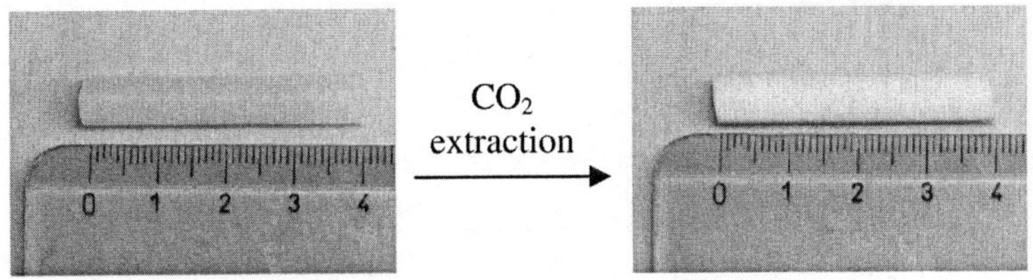

Figure 1. Photographs of a piece of sPS gel prepared in toluene at C_{pol} = 0.10 g/g, before and after total solvent extraction with supercritical CO_2.

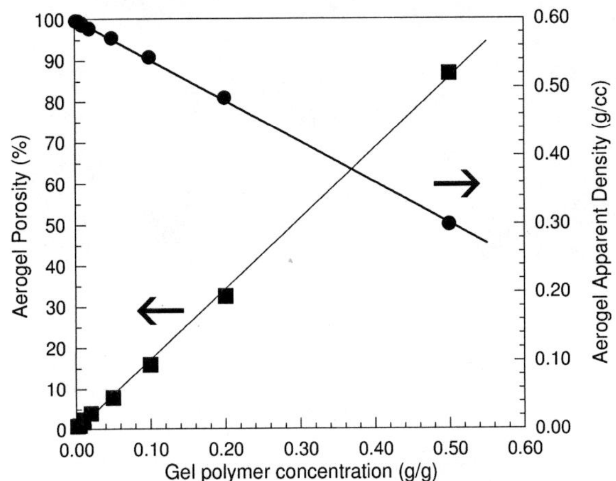

Figure 2. Variation of the apparent density and porosity of aerogels obtained from gels prepared in toluene at different polymer concentrations.

It has been observed that different crystalline phases and morphologies can be obtained in sPS aerogels depending on the starting gel type (type I or II) or on the supercritical CO_2 extraction temperature (*44–46*) and aerogels with the densely-packed γ (*46*) and β (*44, 46*) forms or with the microporous δ (*44, 45*) and ε (*46*) forms can be obtained.

In the following sections, different aspects relative to the structure, the morphology and the sorption properties of syndiotactic polystyrene aerogels will be discussed. In particular the sorption properties of nitrogen at 77 K and of organic compounds from vapour and diluted aqueous solutions by sPS aerogels with different porosities and crystalline forms will be presented.

Structure of sPS Aerogels

δ-Form Aerogels

Typical X-ray diffraction patterns of aerogels obtained after CO_2 extraction at 40°C of the solvent present in physical gels characterized by a co-crystalline phase are shown in Figure 3.

The presence of strong reflections located at 2θ (CuKα) 8.3°, 13.5°, 16.8°, 20.7°, 23.5° and the absence of a diffraction peak at 2θ ≅ 10.6° indicate that

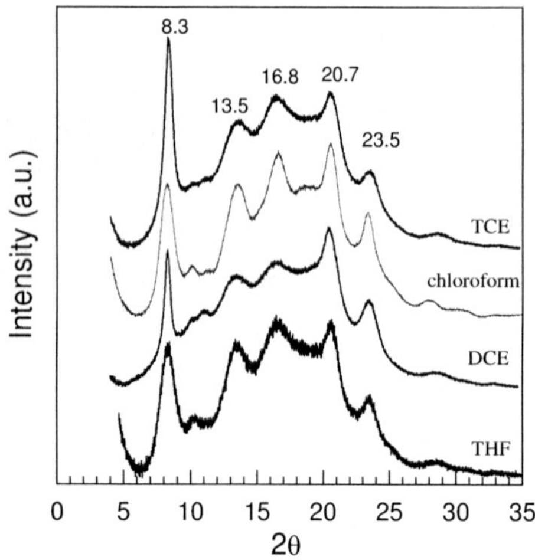

Figure 3. X-ray diffraction patterns of aerogels obtained from gels prepared in trichloroethylene (TCE), chloroform, 1,2-dichloroethane (DCE), and tetrahydrofuran (THF) at C_{pol} = 0.10 g/g.

the CO_2 treatment has extracted the solvent guest molecules also from the co-crystalline phase initially present in the gels thus producing the microporous δ-form (*7*).

Results have shown that sPS aerogels obtained from gels with a co-crystalline phase after supercritical CO_2 treatment present a similar fibrillar morphology (*44, 46, 51*).

This is shown, for instance in Figure 4 by the scanning electron micrographs of aerogels obtained by supercritical CO_2 extraction from gel prepared in TCE and DCE, at C_{pol} = 0.02 g/g.

Thus, for sPS gels characterized by a co-crystalline phase, the corresponding aerogels are formed by semicrystalline nanofibrils with fiber diameters in the range 20-200 nm presenting the microporous crystalline δ-form (as schematically shown in Figure 5).

γ-Form Aerogels

It has been observed that supercritical extraction temperature has an effect on the crystalline structure of sPS aerogels (*46*). The X-ray diffraction pattern and the scanning electron micrograph of an aerogel obtained after CO_2 extraction at 130°C of the solvent present in a gel characterized by a co-crystalline phase are reported in Figure Figure 6.

The X-ray diffraction pattern of the aerogel displays strong reflections at 2θ = 9.4, 10.4, 16 and 20° thus indicating the formation of the γ-form (*5, 6*).

The micrograph shown in Figure 6B clearly shows that after supercritical CO_2 extraction at 130°C the fibrillar morphology is maintained and a γ-form aerogel with nanofibrils having a diameter in the range 30-150 nm is obtained.

It is worth noting that, unlike supercrtical CO_2 extraction, thermal treatments at 130°C of δ aerogels induces a significant decrease of the aerogel porosity (*46*).

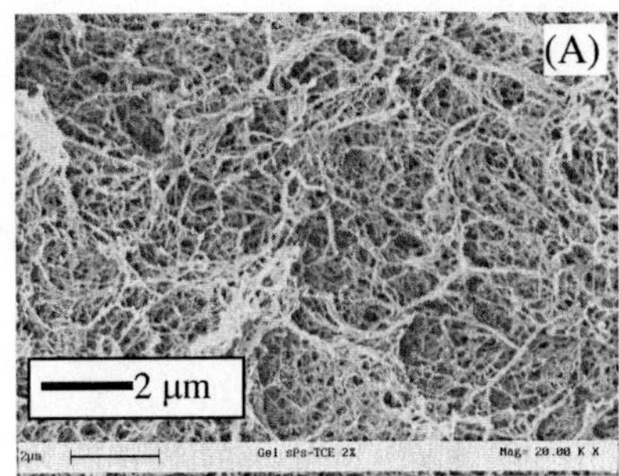

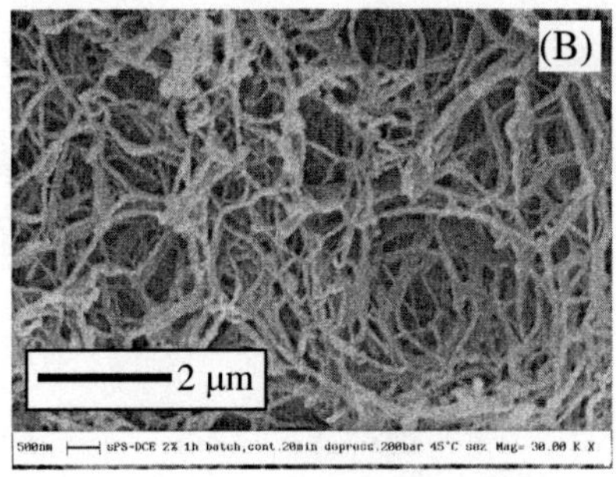

Figure 4. Scanning electron micrographs of aerogels obtained from gels prepared at C_{pol} =0.02 g/g in TCE (A) and DCE (B).

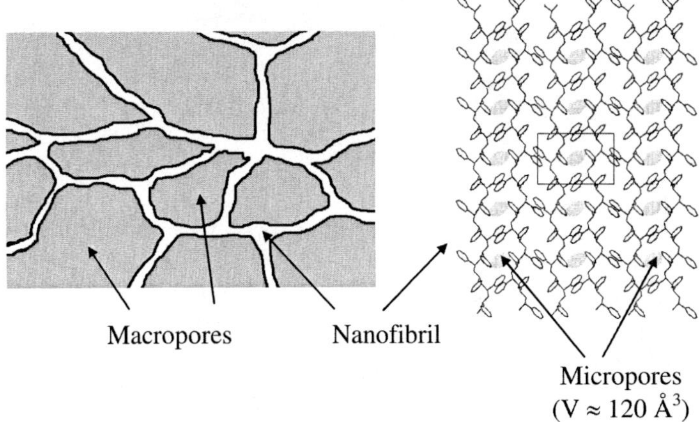

Figure 5. Schematic presentation of the texture (left) and of the crystalline structure (right, the nanocavities are shaded) of δ aerogels obtained from gels characterized by a co-crystalline phase. (C. Daniel, D. Alfano, V. Venditto, S. Cardea, E. Reverchon, D. Larobina, G. Mensitieri, G. Guerra: Aerogels with a Microporous Crystalline Host Phase. Advanced Materials. 2005. 17. 1515-1518. Copyright Wiley-VCH Verlag GmbH & Co. KGaA. Reproduced with permission.)

β-Form Aerogels

High porosity monolithic β-form aerogels can be obtained by CO_2 treatment at 40°C of a type II gel characterized by the β-form (*44*) or by CO_2 treatment at 150°C of a type I gel characterized by a co-crystalline phase (*46*).

In Figure 7 are reported the x-ray diffraction patterns (curves a and b of Figure 7A) and the scanning electron micrographs of aerogels obtained with both procedures (Figures 7B and 7C).

The X-ray diffraction pattern of the aerogel obtained by CO_2 treatment at 40°C of the sPS/chlorotetradecane gel (curve a of Figure 7A) displays strong reflections at 2θ = 6.1°, 10.4°, 12.3°, 13.6°, 18.5° and 20.2° thus indicating the preservation within the aerogel of the orthorhombic β-form (*3, 4*), that was already present in the starting gel (*38, 41*). The x-ray diffraction pattern of the aerogel obtained by

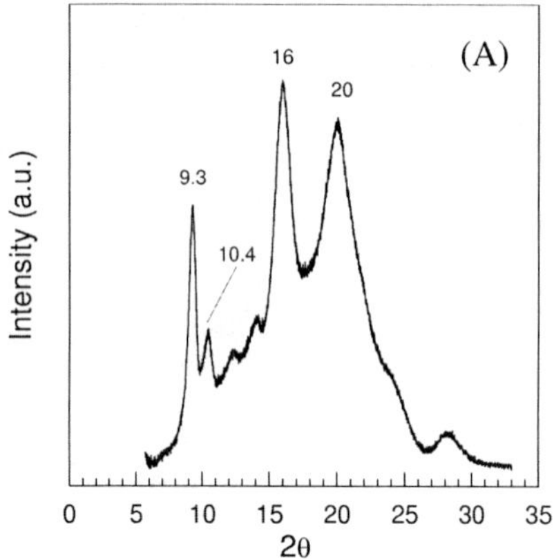

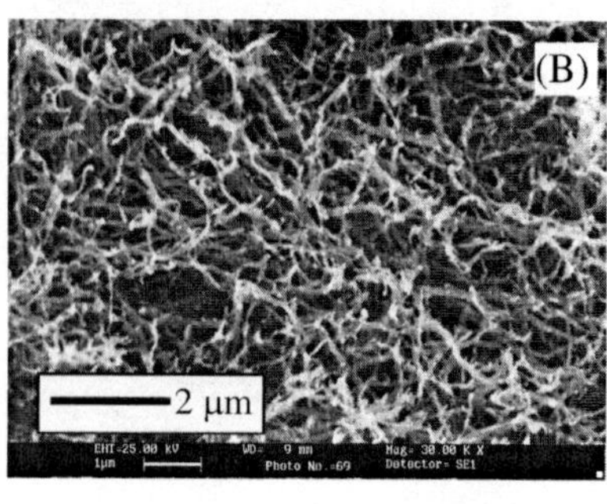

Figure 6. X-ray diffraction pattern (A) and scanning electron micrograph (B) of a sPS aerogel obtained from a gel prepared in DCE at C_{pol} = 0.05 g/g, by supercritical CO_2 extraction at 130°C.

CO_2 treatment at 150°C of a sPS/DCE gel (curve b of Figure 7B) shows also the presence of the β-form in the aerogel.

However, if both procedures lead to β-form aerogels, the micrographs reported in Figures 7B and 7C show that the morphology of the aerogels are totally different. The aerogel obtained from sPS/CTD gel shows interconnected lamellar crystals with a thickness in the range 200-400 nm (Figure 7B) while the aerogel obtained from the sPS/DCE gel shows fibrillar network morphology (Figure 7C) similar to the one observed for δ-form and γ-form aerogels (see Figures 4 and 6B).

ε-Form Aerogels

The second microporous crystalline phase of sPS, named ε (*8*, *9*), can be obtained by treatment with liquid chloroform of γ-form samples (film or powder) followed by chloroform removal, by volatile guests of sPS, like acetone or acetonitrile or CO_2 (*8*, *9*).

Monolithic aerogels characterized by the ε-form and a porosity up to 90% have been also prepared by immersion in chloroform of γ aerogel obtained by supercritical CO_2 extraction at 130°C of a gel with a co-crystalline phase (see above) followed by chloroform removal with supercritical CO_2 at 40°C (*46*).

Typical x-ray diffraction pattern and electron micrograph of the thus obtained aerogel are reported in Figure 8A and 8B, respectively.

The x-ray diffraction pattern of Figure 8A displays strong reflections at 2θ = 6.9, 8.1, 13.7, 16.2, 20.5, 23.5, and 28.5° indicating the formation of the ε crystalline form (*9*) which is characterized by channel-shaped cavities crossing the unit cells along the c axis. The scanning electron micrograph shows that the fibrillar morphology of the starting γ aerogel (see Figure 6B) is substantially modified by the chloroform treatment, so that the obtained ε aerogel presents a sort of open pore morphology, with pores with diameters in the range 0.5-2μm

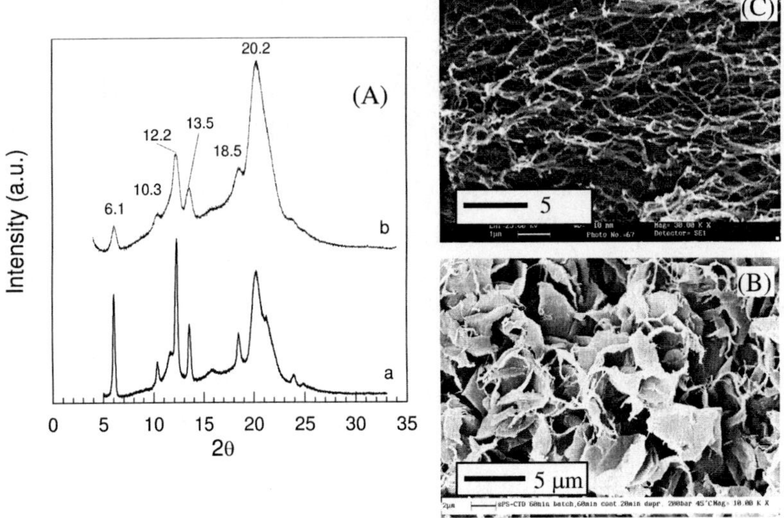

Figure 7. (A) X-ray diffraction patterns of an aerogel obtained from a type II gel (sPS/chlorotetradecane, C_{pol} = 0.10 g/g) by CO_2 extraction at 40°C (curve a) and of an aerogel obtained from a type I gel (sPS/1,2-dichloroethane, C_{pol} = 0.05 g/g) by CO_2 extraction at 150°C (curve b); (B) scanning electron micrograph of the aerogel of curve a; (C) scanning electron micrograph of the aerogel of curve b.

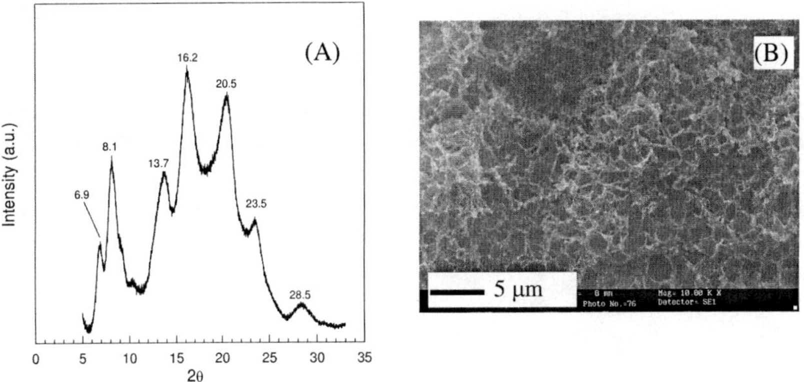

Figure 8. X-ray diffraction pattern (A) and scanning electron micrograph (B) of a sPS aerogel (P = 89%) obtained from a γ-form aerogel (P = 93%), after immersion in chloroform followed by supercritical CO_2 extraction at 40°C.

Thus, ε aerogels are characterized by macropores in the range 0.5-2 μm and a polymer framework presenting the channel-shaped microcavities of the crystalline ε-form (as schematically shown in Figure 9).

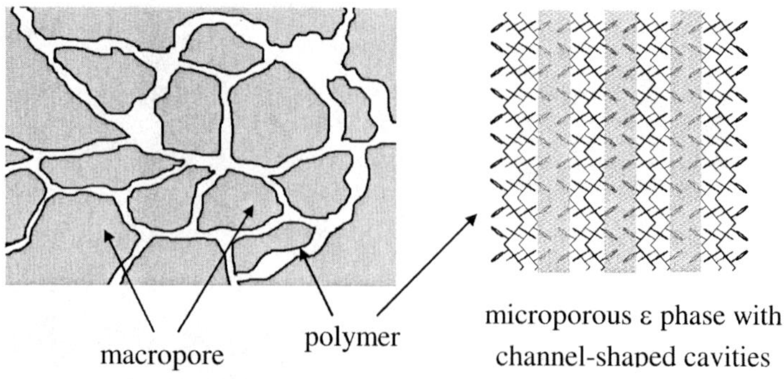

Figure 9. Schematic representation of the texture (left) and of the microporous crystalline structure (right, the nanocavities are shaded) of ε aerogels.

Sorption Properties

Nitrogen Sorption

Following a standard procedure for porosity evaluation of powders (*52*), nitrogen sorption experiments at low temperature (77 K) have been conducted with sPS aerogels with different crystalline structures and porosities (*45*, *46*).

Nitrogen sorption isotherms in the range 0-0.30 p/p_0 of δ aerogels with a porosity P = 91% and P = 98.5%, of an ε aerogel (P = 89%) and a β aerogel (P = 91%) are compared with the sorption isotherms of δ and γ powders (70-100 mesh size, empty symbols) (*20*) in Figure 10, where the sorption is expressed as cm^3 of nitrogen at normal conditions (1 atm, 0°C) per gram of polymer.

The sorption curves of Figure 10 and relevant data derived from them (reported in Table I) show that the sorption capacity is markedly larger for samples presenting the crystalline nanocavities of the δ and ε forms and it also markedly increases with the aerogel porosity.

As discussed in detail in reference (*20*), the nitrogen sorption observed with dense sPS samples (powder or films) is nearly entirely due to its inclusion as guest into the nanocavities of the δ crystalline phase.

With reference to the sorption curves of Figure 10, it is also worth noting that the sum of nitrogen uptake from δ powder (empty circles, corresponding to absorption only in the crystalline nanocavities (*20*)) and β aerogel (filled triangles) is comparable with the nitrogen uptake from the δ aerogel of equal porosity (filled squares). Nitrogen uptake from β aerogels, due to the negligible empty volume (*20*) and guest uptake (*53*, *54*) from its high density (1.078 g/cm^3) β crystalline phase (*3*, *4*), is entirely due to adsorption on the amorphous porosity. This clearly indicates that δ aerogels, beside the usual nitrogen absorption into the crystalline nanocavities, permit a substantial nitrogen adsorption on the large surface area of the amorphous sample pores.

The curves of Figure 10 also show that the sorption capacity of ε aerogels (crosses) which are also characterized by the presence of crystalline nanocavities present a higher sorption capacity than β aerogels.

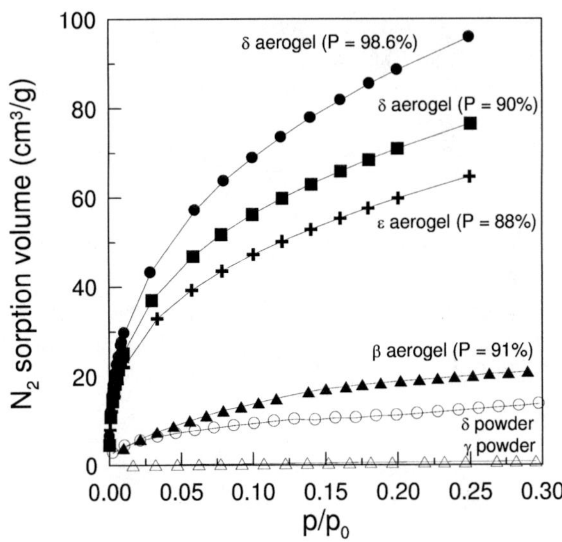

Figure 10. Nitrogen sorption isotherms in the range 0-0.30 p/p_0 of δ aerogels (P=91% and 98.6%), ε aerogel (P = 88%), β aerogel (P=91%), δ powder and γ powder at 77K. The sorption is expressed as cm^3 of nitrogen at normal conditions (1 atm, 273K) per gram of polymer.

Table I. Nitrogen sorption and BET values, as derived by isotherm curves of Figure 10

sPS sample	*N_2 sorption at p/p°=0.1 (cm^3/g)*	*BET[a] (m^2/g)*
γ powder	0.3	4
δ powder	9	43
β aerogel (P= 91%)	14	70
ε aerogel (P = 90%)	47	230
δ aerogel (P = 90%)	56	270
δ aerogel (P = 98.5) %)	69	348

[a] as evaluated in the range 0.05-0.30 p/p°.

It is also worth noting that the increments of nitrogen uptake, which are observed when we compare δ powder and δ aerogels, increase with aerogel porosity. This confirms that the increment of nitrogen uptake of aerogels with respect to the powder is due to adsorption in the amorphous pores.

Sorption of VOC from Vapor Phase

The sorption capacity of VOC from vapor phase in sPS aerogels has been investigated (*44*) and chloroform vapor sorption kinetic curves carried out at partial pressure $P_{chloroform}$= 667 Pa and T = 56°C are compared in Figure 11 with the sorption curve obtained with a δ-form film (8.5 μm thickness).

The data clearly show that the δ aerogel possesses a much higher sorption capacity than the β aerogel (2.6 vs. 0.33 $g_{chloroform}/100\ g_{aerogel}$). We can also observe that the sorption capacity of the δ-aerogel is also slightly higher than for the δ-form film (likely due to a slightly higher crystallinity) but the sorption kinetics of the aerogel is much faster than for the film, although the macroscopic dimensions of the aerogel are larger than the film thickness. This is related to the effective dimension to be considered when analysing chloroform diffusion into the aerogel, that is the fibril diameter rather than the external dimensions

Thus, for low VOC vapor pressures, the δ aerogel present the high sorption capacity characteristic for sPS δ form samples (due to the sorption of the organic molecules as isolated guests of the host microspores crystalline phase) associated with the high sorption kinetics typical for aerogels (due to the high porosity and the large pore size.

Sorption of VOC from Dilute Aqueous Solutions

The sorption of 1,2-dichloroethane (DCE) from diluted aqueous solutions has been investigated for several sPS aerogels (*45*, *46*), and DCE equilibrium uptakes from δ aerogels (P = 90%), ε aerogels (P = 88%), and γ aerogels (P = 90%) are compared in Figure 12 with the equilibrium sorption uptakes of DCE from δ powder (*45*) and activated carbon (*55*).

We can observe that the VOC uptake of δ-form powder and aerogels are almost identical. Particularly relevant are the results relative to the VOC uptake from the most diluted aqueous solution (1 ppm) showing that the sorption capacity in δ aerogel is larger than 5 $g_{DCE}/100g_{polymer}$, i.e. leading to a concentration increase of 50000 times.

Moreover, it is clearly apparent that the DCE uptake from the γ aerogel is always much lower than for the δ-form samples and becomes negligible for dilute aqueous solutions. This clearly indicates that the DCE uptake occurs essentially only in the crystalline microporous phase.

It is also worth noting that the uptake of the δ aerogel is more than twice higher than for activated carbon.

With regard to ε aerogels, Figure 12 shows that the equilibrium DCE uptake from the ε aerogel is lower than for the δ-form aerogel but much larger than for the γ aerogel. This clearly indicates that, as for δ aerogels, the DCE uptake occurs essentially only in the channel-shaped micropores of the ε-form. It is also worth noting that the DCE uptake, from highly diluted aqueous solutions, for ε aerogels remains also higher than for activated carbons.

DCE sorption kinetics of δ aerogels with a porosity P = 98.5%, 90%, and 80% from 100 ppm aqueous solutions (*45*), are compared in Figure 13A while DCE desorption experiments at room temperature in air after equilibrium absorption from 100 ppm aqueous solution are shown in Figure 13B. For the sake of comparison, the desorption kinetics for an unoriented film, having a thickness of nearly 40μm (*24*), is also reported as inset in Figure 13B.

In Figure 13A, the DCE uptake is reported versus square root of time divided by the aerogel macroscopic thickness, to put in evidence Fick's sorption behavior

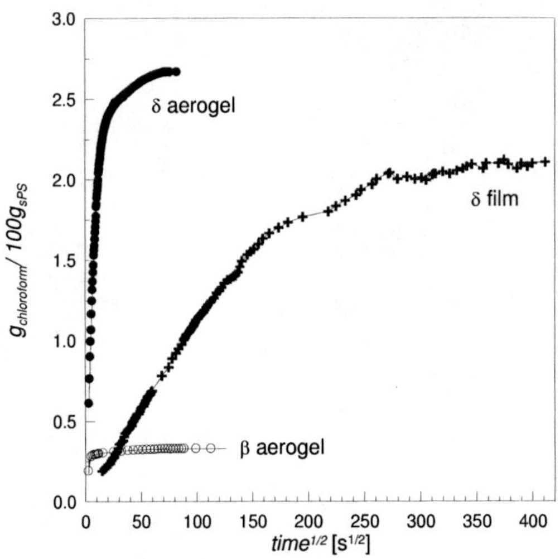

Figure 11. Chloroform vapor sorption kinetics obtained at $P_{chloroform}$= 667 Pa and T = 56°C for the δ (•) and β (○) aerogels (cylinder shape with 10 mm length and 5 mm diameter) and for a the δ-form film (8.5 μm thickness) (+). (C. Daniel, D. Alfano, V. Venditto, S. Cardea, E. Reverchon, D. Larobina, G. Mensitieri, G. Guerra: Aerogels with a Microporous Crystalline Host Phase. Advanced Materials. 2005. 17. 1515-1518. Copyright Wiley-VCH Verlag GmbH & Co. KGaA. Reproduced with permission.)

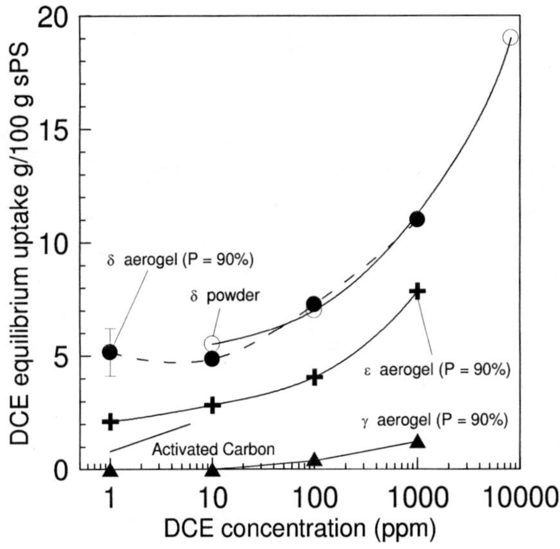

Figure 12. Equilibrium DCE sorption at room temperature by δ aerogels (P = 90 %, filled circles), ε aerogels (P = 90 %, crosses), γ aerogels (P=90%, triangles), δ powder (open circles) and activated carbon (thin line).

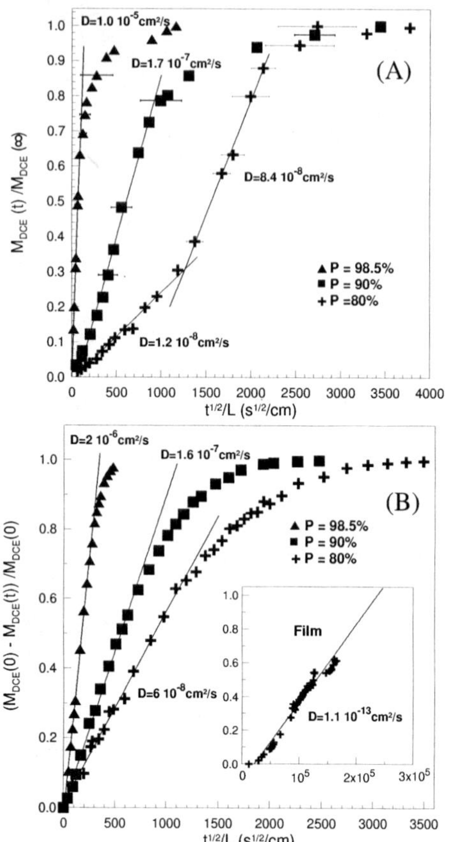

Figure 13. Sorption kinetics at room temperature of DCE from 100 ppm DCE aqueous solutions (A) and desorption kinetics of DCE at room temperature, after equilibrium sorption uptake from 100 ppm DCE aqueous solution (B), for δ aerogels with a porosity P = 98.5% (▲), P = 90% (■), and P = 80% (+). The desorption kinetics of an unoriented δ film is reported as inset in Figure 13B.

and the apparent diffusivities (D) as evaluated from line slopes have been collected in Table II along with the diffusivity constants of an unoriented δ film.

We can clearly observe a significant increase of the apparent diffusivity with the aerogel porosity (i.e. specific surface area). In particular, for a 100 ppm DCE aqueous solution, the diffusivity constant of a 98.5% porosity aerogel is c.a. 60 times larger than for a 90% porosity aerogel, and c.a. 120 times larger than for a 80% aerogel.

The diffusivity constants of the aerogels are much higher than those obtained for films. In particular, the diffusivity of a δ-aerogel with P = 98.5% in a 100 ppm DCE aqueous solution is nearly 7 orders of magnitude larger than the diffusivity constant obtained for δ-form films (*24*). Of course, the higher apparent diffusivities observed with aerogels are due to their higher specific surface area.

Table II. Apparent diffusivity constants at room temperature of DCE in sPS samples presenting the microporous δ crystalline phase. Data for the absorption and desorption from diluted aqueous solutions is shown

δ-form samples of sPS	*D[cm²/s] (absorption)*	*D[cm²/s] (desorption)*
Unoriented film[a]	8.6×10^{-13}	1.1×10^{-13}
Aerogel, P=80%	8.4×10^{-8} (100ppm)	6.0×10^{-8}
Aerogel, P=90%	1.7×10^{-7} (100ppm)	1.6×10^{-7}
Aerogel, P=98.5%	1.0×10^{-5} (100ppm)	2.0×10^{-6}

[a] data from ref. (*24*), which refer to sorption of DCE from 50 ppm aqueous solutions into films with a thickness of c.a. 40 μm.

As for the desorption experiments, the diffusivities are smaller than those of the corresponding sorption experiments, as already observed for δ-form films (*22*). It is also worth noting than, after 3000 h of desorption, nearly 2 wt % of DCE (nearly 40% of the initial content) is still present in the polymer film while complete DCE desorption occurs in c.a. 50 and 10 hours for aerogels with 80% and 98.5% porosity, respectively.

The sorption from dilute aqueous solutions of large molecules being unsuitable to form a co-crystal structure by sorption into the δ-form are been also investigated (*46*) and the FTIR spectra of a δ aerogel (P = 90%) and a ε aerogel (P=88%) before (curves a and c, respectively) and after equilibrium sorption from a 12 ppm 4-(dimethyl-amino)-cinnamaldehyde (DMACA) aqueous solution (curves b and d, respectively) are reported in Figure 14.

Sorption of Long-Large Molecules from Dilute Aqueous Solutions

We can observe that for the δ-form aerogel the FTIR spectra before (curve a in Figure 14) and after immersion in the aqueous solution of DMACA (curve b in Figure 14) are nearly identical. This indicates that DMACA sorption from the aqueous solution for the aerogel with the microporous δ-form is negligible, as for the aerogels with dense crystalline phases (β or γ). Conversely, absorption peaks located at 1674, 1526, 1131, and 1116 cm^{-1} (curve d in Figure 14) clearly indicate a large uptake of DMACA in the ε-form aerogel. For instance, the DMACA uptake in the ε-form aerogel from 12 ppm aqueous solution is close to 3 wt%, as evaluated on the basis of thermogravimetric measurements.

The negligible sorption of the long and bulky guests from β, γ and δ aerogels having similar porosity and morphology clearly indicates that, for low activity, DMACA molecules are absorbed essentially only in the channel-shaped cavities of the ε-form crystalline phase rather than in the amorphous macropores of the aerogel. The absence of sorption in δ-form aerogels is due to the molecular volume of DMACA molecules which are, unlike DCE, too bulky to enter in the microporous cavities of the δ crystalline phase.

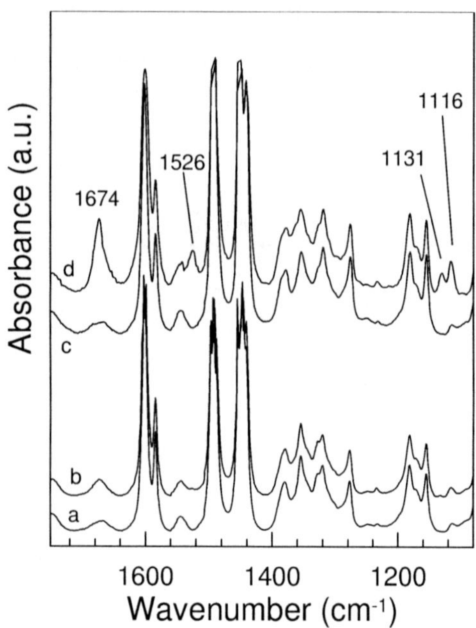

Figure 14. FTIR spectra in the wavenumber range 1080-1750 cm⁻¹ for a δ (a, b) and a ε (c, d) aerogel, before (a, c) and after equilibrium sorption from a 12 ppm DMACA aqueous solution (b, d).

Concluding Remarks

In this contribution different aspects relative to the structure, the morphology and the sorption properties of syndiotactic polystyrene aerogels have been presented.

High porosity monolithic aerogels with fibrillar morphology and the densely-packed γ and β forms are obtained by high temperature CO_2 extraction of type I gels (i.e. characterized by a co-crystalline phase) at 130°C and 150°C, respectively. On the other hand, β aerogels with lamellar morphology are obtained by CO_2 extraction of type II gels (i.e. characterized by the β-form) at 40°C.

High porosity monolithic aerogels characterized both by amorphous macropores and crystalline micropores can be also obtained.

Supercritical CO_2 extraction at 40°C of type I gels leads aerogels formed by semicrystalline nanofibrils (fibril diameter range between 100-200 nm) exhibiting the microporous crystalline δ-form. Treatments of γ-form aerogels with chloroform, followed by solvent extraction by supercritical CO_2 at 40°C, lead to the formation of high porosity monolithic aerogels with the microporous crystalline ε-form.

Sorption of organic molecules from aqueous diluted solutions is high and fast by sPS aerogels, only when a microporous crystalline phase (δ or ε) is present. Moreover, the guest uptake depends strongly on the guest molar volume.

For small molecules such as DCE, molecules can be efficiently absorbed by both δ and ε aerogels, although the guest uptake is higher for δ aerogels. The VOC sorption for δ-aerogels is similar to those observed for other δ-form samples and it is independent of aerogel porosity. These results indicate that the organic guest uptake from diluted aqueous solutions occurs essentially only in the crystalline nanocavities. However sorption and desorption experiments of DCE guest molecules show that the kinetics strongly depend on the aerogel porosity and the use of δ aerogels can result in an apparent increase in the guest diffusivity of several orders of magnitude (up to 7!), with respect to δ-form films.

Conversely, for long molecules, only ε aerogels can efficiently sorb the molecules from diluted aqueous solutions. For these long guests,the channel-shaped cavities of the sPS ε-form play a key role in the aerogel sorption capacity.

These results show that δ -form sPS aerogels present high VOC sorption capacity and fast kinetics as well as convenient handling characteristics. Moreover, by increasing their porosity it is possible to substantially increase their surface area and sorption kinetics. This makes these materials particularly suitable as sorption medium to remove traces of small pollutants from water and moist air while for long organic pollutant molecules, results show that ε-form aerogels are suitable for detection and removal of these molecules.

The low-temperature nitrogen uptake from δ and ε aerogels is much higher than from other sPS samples and strongly increases with aerogel porosity. These phenomena indicate that nitrogen absorption in the crystalline nanocavities is additive with respect to nitrogen adsorption from the amorphous porosity. The high sorption capacity of nitrogen at 77 K makes microporous sPS aerogels potentially interesting for gas storage applications.

References

1. De Rosa, C.; Guerra, G.; Petraccone, V.; Corradini, P. *Polym. J.* **1991**, *23*, 1435–1442.
2. Cartier, L.; Okihara, T.; Lotz, B. *Macromolecules* **1998**, *31*, 3303–3310.
3. De Rosa, C.; Rapacciuolo, M.; Guerra, G.; Petraccone, B.; Corradini, P. *Polymer* **1992**, *33*, 1423–1428.
4. Chatani, Y.; Shimane, Y.; Ijitsu, T.; Yukinari, T. *Polymer* **1993**, *34*, 1625–1629.
5. Immirzi, A.; De Candia, F.; Iannelli, P.; Vittoria, V.; Zambelli, A. *Makromol. Chem., Rapid Commun.* **1988**, *9*, 761–764.
6. Rizzo, P.; Lamberti, M.; Albunia, A. R.; Ruiz de Ballesteros, O.; Guerra, G. *Macromolecules* **2002**, *35*, 5854–5860.
7. De Rosa, C.; Guerra, G.; Petraccone, V.; Pirozzi, B. *Macromolecules* **1997**, *30*, 4147–4152.
8. Rizzo, P.; Daniel, C.; De Girolamo Del Mauro, A; Guerra, G. *Chem. Mater.* **2007**, *19*, 3864–3866.
9. Petraccone, V.; Ruiz de Ballesteros, O.; Tarallo, O.; Rizzo, P.; Guerra, G. *Chem. Mater.* **2008**, *20*, 3663–3668.

10. Chatani, Y.; Shimane, Y.; Inagaki, T.; Iijtsu, T.; Yukimori, T.; Shikuma, H. *Polymer* **1993**, *34*, 1620–1624.
11. Chatani, Y.; Inagaki, T.; Shimane, Y.; Shikuma, H. *Polymer* **1993**, *34*, 4841–4845.
12. De Rosa, C.; Rizzo, P.; Ruiz de Ballesteros, O.; Petraccone, V.; Guerra, G. *Polymer* **1999**, *40*, 2103–2110.
13. Tarallo, O.; Petraccone, V. *Macromol. Chem. Phys.* **2004**, *205*, 1351–1360.
14. Tarallo, O.; Petraccone, V. *Macromol. Chem. Phys.* **2005**, *206*, 672–679.
15. Kaneko, F.; Uda, Y.; Kajivara, A.; Tanigaki, N *Macromol. Rapid Commun.* **2006**, *27*, 1643–1647.
16. Petraccone, V.; Tarallo, O.; Venditto, V.; Guerra, G. *Macromolecules* **2005**, *38*, 6965–6971.
17. Tarallo, O.; Petraccone, V.; Venditto, V.; Guerra, G. *Polymer* **2006**, *47*, 2402–2410.
18. Galdi, N.; Albunia, A. R.; Oliva, L.; Guerra, G. *Macromolecules* **2006**, *39*, 9171–9176.
19. Milano, G.; Venditto, V.; Guerra, G.; Cavallo, L.; Ciambelli, P.; Sannino, D. *Chem. Mater.* **2001**, *13*, 1506–1511.
20. Manfredi, C.; Del Nobile, M. A.; Mensitieri, G.; Guerra, G.; Rapacciuolo, M. *J. Polym. Sci., Polym. Phys. Ed.* **1997**, *35*, 133–140.
21. Musto, P.; Mensitieri, G.; Cotugno, S.; Guerra, G.; Venditto, V. *Macromolecules* **2002**, *35*, 2296–2304.
22. Mahesh, K. P. O.; Tsujita, Y.; Yoshimizu, H.; Okamoto, S.; Mohan, D. *J. Polym. Sci., Part B: Polym. Phys.* **2005**, *43*, 2380–2387.
23. Venditto, V.; De Girolamo Del Mauro, A.; Mensitieri, G.; Milano, G.; Musto, P.; Rizzo, P.; Guerra, G. *Chem. Mater.* **2006**, *18*, 2205–2210.
24. Mensitieri, G.; Venditto, V.; Guerra, G. *Sens. Actuators, B* **2003**, *92*, 255–261.
25. Giordano, M.; Russo, M.; Cusano, A.; Cutolo, A.; Mensitieri, G.; Nicolais, L. *Appl. Phys. Lett.* **2004**, *85*, 5349–5351.
26. Giordano, M.; Russo, M.; Cusano, A.; Mensitieri, G.; Guerra, G. *Sens. Actuators, B* **2005**, *109*, 177–184.
27. Cusano, A.; Pilla, P.; Contessa, L.; Iadicicco, A.; Campopiano, S.; Cutolo, A.; Giordano, M.; Guerra, G. *Appl. Phys. Lett.* **2005**, *87*, 234105/1–234105/3.
28. Kobayashi, M.; Nakaoki, T.; Ishihara, N. *Macromolecules* **1990**, *23*, 78–83.
29. Kobayashi, M.; Kosaza, T. *Appl. Spectrosc.* **1993**, *9*, 1417–1424.
30. Deberdt, F.; Berghmans, H. *Polymer* **1993**, *34*, 2192–2201.
31. Daniel, C.; Dammer, C.; Guenet, J. M. *Polymer* **1994**, *35*, 4243–4246.
32. Kobayashi, M.; Yoshioka, T.; Kozasa, T.; Tashiro, K.; Suzuki, J.-I.; Funahashi, S.; Izumi, Y. *Macromolecules* **1994**, *27*, 1349–1354.
33. Daniel, C.; Deluca, M. D.; Guenet, J. M.; Brulet, A.; Menelle, A. *Polymer* **1996**, *37*, 1273–1280.
34. Roels, T.; Deberdt, F.; Berghmans, H. *Progr. Colloid Polym. Sci.* **1996**, *102*, 82–85.
35. Daniel, C.; Menelle, A.; Brulet, A; Guenet, J.-M. *Polymer* **1997**, *38*, 4193–4199.

36. Rastogi, S.; Goossens, J. G. P.; Lemstra, P. J. *Macromolecules* **1998**, *31*, 2983–2998.
37. Li, Y.; Xue, G. *Macromol. Rapid Commun.* **1998**, *19*, 549–552.
38. Daniel, C.; Musto, P.; Guerra, G. *Macromolecules* **2002**, *35*, 2243–2251.
39. Van Hooy-Corstjens, C. S. J.; Magusin, P. C. M.; Rastogi, S.; Lemstra, P. J. *Macromolecules* **2002**, *35*, 6630–6637.
40. Daniel, C.; Alfano, D.; Guerra, G.; Musto, P. *Macromolecules* **2003**, *36*, 1713–1716.
41. Daniel, C.; Alfano, D.; Guerra, G.; Musto, P. *Macromolecules* **2003**, *36*, 5742–5750.
42. Shimizu, H.; Wakayama, T.; Wada, R.; Okabe, M.; Tanaka, F. *Polym. J.* **2005**, *37*, 294–298.
43. Malik, S.; Rochas, C.; Guenet, J. M. *Macromolecules* **2006**, *39*, 1000–1007.
44. Daniel, C.; Alfano, D.; Venditto, V.; Cardea, S.; Reverchon, E.; Larobina, D.; Mensitieri, G.; Guerra, G. *Adv. Mater.* **2005**, *17*, 1515–1518.
45. Daniel, C.; Sannino, D.; Guerra, G. *Chem. Mater.* **2008**, *20*, 577–582.
46. Daniel, C.; Giudice, S.; Guerra, G *Chem. Mater.* **2009**, *21*, 1028–1034.
47. Malik, S.; Rochas, C.; Guenet, J. M. *Macromolecules* **2005**, *38*, 4888–4893.
48. Malik, S.; Roizard, D.; Guenet, J. M. *Macromolecules* **2006**, *39*, 5957–5959.
49. Gesser, H. D.; Goswami, P. C. *Chem. Rev.* **1989**, *89*, 765–788.
50. Ko, E. I. In *Kirk-Othmer Encyclopedia of Chemical Technology*, 4th ed.; John Wiley & Sons: 1998; Aerogels Supplement Volume, pp 1−22.
51. Daniel, C; Avallone, A.; Guerra, G. *Macromolecules* **2006**, *30*, 7578–7582.
52. Gregg, S. J.; Sing, K. S. W. In *Adsorption, Surface Area, and Porosity*; Academic Press: London, 1982.
53. Vittoria, V.; Russo, R.; de Candia, F. *J. Macromol. Sci. Phys.* **1989**, *B28*, 419–431.
54. Rapacciuolo, M.; De Rosa, C.; Guerra, G.; Mensitieri, G.; Apicella, A; Del Nobile, M. A. *J. Mater. Sci. Lett.* **1991**, *10*, 1084–1087.
55. Stanzel, M. H. *Chem. Eng. Progr.* **1993**, *89* (4), 36–43.

Chapter 11

Structure and Properties of Dangling Chain Poly(urethane-isocyanurate) Model Networks†

J. Budinski-Simendić,*,[1] M. Špirkova,[2] J. Pavličević,[1] J. Šomvarsky,[3] K. Mészáros Szécsényi,[4] and K. Dušek[2]

[1]University of Novi Sad, Faculty of Technology, Novi Sad, Serbia
[2]Institute of Macromolecular Chemistry ASCR, Prague, Czech Republic
[3]Charles University, Prague, Czech Republic
[4]Faculty of Sciences, Chemistry Department, Novi Sad, Serbia
***jarka@uns.ac.rs**

†Dedicated to Professor F. E. Karasz on the occasion of his 75th birthday

A new trend in design of functional crosslinked systems is their preparation from specially designed precursors. A model endlinked network is the one in which all the precursor polymers become elastic chains. In real networks, however, imperfections cause the elastic properties of the networks to diminish. These imperfections include dangling chains, which are attached to the network at one end only, at equilibrium deformation, they do not contribute to the retroactive force of the network. In this project we prepared model networks with isocyanurate(hexahydro-1,3,5-triazin-2,4,6-trion) ring as a crosslink points based on 2,4-Tolylene diisocyanate and α,ω, dihidroxypoly (oxypropylene)diols by tertiary amine catalytic cyclotrimerisation in a bulk. The controlled content of dangling chains was obtained by partial substitution of polyol with monool component in synthesis of telechelic diisocyanates. The properties of networks were estimated by multiple extractions in organic solvent, differential scanning calorimetry, dynamic-mechanical analysis, photoelastical measurements. The network topology was calculated using the stochastic theory of branching processes. The thermal stability of some extracted and non-extracted samples was characterized by TGA and simultaneous TG/DSC thermal analyzer SDT Q600 TA Instruments.

1. Introduction

Some type of permanent structure is necessary to form a coherent solid and prevent liquidlike behaviour of elastomer molecules. This requirement is met by incorporating a small number of intermolecular chemical bonds (crosslinks) to make a loose three-dimensional molecular network (*1*). Such crosslinks are generally assumed to form in the most probable positions, so that the long sections of molecules between them have the same spectrum of end-to-end lengths as a similar set of uncrosslinked molecules would have. Even through macroscopically homogeneous, a model network differs from an ideal network by the presence of structural defects. However, endlinking processes provide some control over these defects. The possible arrangements of chains leading to microscopic defects (a) represents unreacted functionalization yielding dangling chains; (b) corresponds to loops that are formed if a chain is linked by its two ends to the same cross-link; (c) shows a multiple connection between two branch points; (d) corresponds to permanently trapped entanglements between two adjacent cross-links (*2*, *3*). However, in many cases the number of potential cross-links is limited, e.g., the primary chains are composed of more than one type of structural units. Block copolymers with glassy or crystalline domains and ionomer networks can serve as other examples (*4*). Moreover, in many physical thermo-reversible gels, the chains are connected by junction zones of high and, in general, unknown functionalities. Verification of theories of rubber elasticity as well as deeper understanding of relations between the structure and mechanical and other physical properties of polymer networks require knowledge of detailed network topology (*5*).

Typical building units used in theory of branching processes (TBP) are monomeric units or their parts, but sometimes larger fragments are formed during the network formation process, or larger pre-prepared molecules with a given nonrandom degree of polymerization/composition distribution enter the cross-linking reaction (*6*). Such units (chains) should be taken as building units in order to keep the precision of structure generation. In such a case, substructures of sol and gel are not composed of the whole building units but only of parts of them. Thus, the structure of such objects depends on the internal structure of the building units, and additional specific considerations are necessary in order to characterize these objects. The various substructures in a gel can be defined as follows (*7*): (a) an elastically active cross-link (EAC) is a junction issuing three or more bonds with infinite continuation, (b) an elastically active network chain (EANC) is a sequence of units issuing one or two bonds with infinite continuation, (d) an elastically active backbone chain (BC) is a sequence of units issuing two bonds with infinite continuation (an EANC without dangling chains) and (e) a dangling chain (DC) is a sequence of units issuing one bond with infinite continuation. Compared to conventional random crosslinking, the end-linking method has a great advantage. One can evaluate the structural parameters such as the number of elastic chains, cross-links and dangling chains in the resulting networks using data on the amount of each reactant and nonextractable components (gel) by theory of branching processes. A new trend in design of functional cross-linked systems is their preparation from specially designed precursors (*8*). These precursors are oligomeric or polymeric compounds or their blends with functional groups attached to a designed backbone which may contain several branch points. Usually, these precursors are cross-linked with another compound (cross-linker) or precursor having other groups. The gelation and network build-up depend on the number of functional groups per precursor, their reactivity, sequence of possible several reaction paths and reaction regime. The variable characterizing the advancement of the network build-up is conversion of functional groups or reaction time. The use of conversion as process variable has the advantage that it is in many respects universal compared to time. The dependencies of structural characteristics on conversion are thus independent of catalyst concentration, in a limited range on temperature, dilution of the system, etc. The chemical composition and architectural arrangements of the backbone determine the general properties important in application, such as build-up of viscosity during crosslinking, pot life, temperature range of glass transition, modulus of elasticity, mechanical ultimate properties.

The basic variables characterizing the network build-up include molecular mass averages before the gel point, the gel point conversion, sol and gel fractions, and effective cross-linking density. This set of information can be obtained from theories of network formation. The main goal of industrial materials research is to formulate procedures that will lead to the design of polymer networks with preset, i.e. specified, properties for different application (*9*). Polyurethane materials, with their wide range of possible types, open the way to manufacturing of an enormous range of materials with different chemical and physical properties. They are one of the most useful commercial classes of polymers which are widely used in both industry and in everyday life (*10*). Polyurethanes have been modified for wide ranging applications such as coatings, adhesives, resins, foams, membranes, synthetic leathers, elastomers and biomedical materials (*11–15*). Despite the possibility of tailoring their properties according to requirements, they suffer a disadvantage of poor stability towards heat. Chemical modification of polyurethane structure by introducing the stable isocyanurate(hexahidro-1,3,5-triazin-2,4,6-trion) rings can improve their thermal properties (*16*, *17*). The ability of an isocyanate to trimerize is strongly dependent on the electronic structure and thus on the electronic environment of the isocyanate group. The double bonds between the nitrogen, carbon and oxygen atoms are polarized, because of the difference in electronegativity. Due to the resonance, averaged over time, the carbon atom is positively charged making it sensitive to an attack by nucleophilic species. Such nucleophilic addition reactions can be catalyzed either by Lewis base (e.g. tertiary amine) by opening the π-system of the isocyanate group or by Lewis acids through an interaction which deforms the π-electron cloud and perturbs its symmetry (e.g. organic metal compounds). According to the proposed mechanisms, the reactivity is strongly dependent on the group connected to the isocyanate group. Electron-withdrawing groups adjacent to the NCO group enhance the reactivity by making the isocyanate-carbon atom more electrophilic. Electron-donating groups, on the other hand, will lower the reactivity. This means that aromatic isocyanates will trimerize significantly faster than aliphatic or cycloaliphatic isocyanates. By replacing hydrogen atoms by more electronegative atoms such as fluorine or chlorine, enhancement of the reactivity of the isocyanate is expected. Steric hindrance of the NCO-group by bulky side groups will undoubtedly influence the reactivity by lowering it through a shielding. A further influence is the rigidity of the connected group, which becomes extremely important for polytrimerization of difunctional isocyanates. The longer space between the isocyanates, the better the polycyclotrimerization will run (*18*). For the network preparation in a bulk, the catalyst system should be molecularly soluble in the network precursor to produce final materials with high optical quality. The complete information about the structure of networks is helpful for a better understanding of the thermal decomposition (*19*). Many applications of PU require structure that can resist a variety of external stresses such as heat and fire among others (*20*).

The formation of polyurethane networks can be accompanied by a number of side reactions owing to a high reactivity of [NCO] groups and this fact considerably complicates the manufacturing of elastomeric materials with defined structure and desired thermal stability. Semsarzadeh at al. studied the reactive extrusion of poly(urethane-isocyanurate) networks in an intermeshing corotating twin-screw extruder (*21*). They also investigated the thermal stabilities, morphologies, and dynamic mechanical properties of reactively extruded poly(urethane-isocyanurate) networks. Pure PU generally have low stability at high temperatures and degradation can occur at temperatures above 180°C, depending on the parameters already discussed. It is well documented in the literature that the chemical modification of the PU backbone through the introduction of heterocyclic structures such as isocyanurate, oxazolidone, imide, triazine or phosphazene are procedures for improving its thermal stability (*22*). In our earlier study, thermal stability of model poly(urethane-isocyanurate) networks was investigated (*23*). It has been found that evaluation of structural parameters for model networks with hexahydro-1,3,5-triazin-2,4,6-trion) ring as a junction points can be calculated by applying the theory of branching processes. The aim of this work has been focused on damping properties and thermal stability of dangling chain polyurethane networks which are suitable for the applications that require heat resistance.

2. Experimental

2.1. Materials

2,4-Tolylene diisocyanate (TDI; Fluka) was purified by distillation. Its purity, 99.9%, was determined by potentiometric titration. Alfa,omega-dihydroxypoly (oxypropylenes) (PPD) of nominal molecular weights 725, 2000, 4000 (Aldrich) and 1200 (Union Carbide) were dried at 50 °C under reduced pressure (ca. 270 Pa) for a time necessary to achieve water contents of several tens of ppm. The ratios of [OH] group from DEGME X=[OH]monool/ [(OH]monool+[OH]diol) were: 0; 0.10; 0.15; 0.20; 0.30, 0.50. Number-average molecular weights, M_n, for all diols, have been determined by vapour-pressure osmometry, water content by coulometry, and the hydroxyl group content by the phenyl isocyanate method (reaction of polyols with excess of phenyl isocyanate, followed by reaction with dibutylamine (DBA) and potentiometric titration of unreacted DBA with HCl in isopropyl alcohol). The number-average functionality, *f*, was calculated from experimentally determined *Mn* and the hydroxyl group content.

2.2. Network Precursor Preparation

A three-neck round-bottom flask equipped with a stir bar, a nitrogen/vacuum adapter was charged with PPD. The system was evacuated to 100 mTorr and degassed by successive cycles of nitrogen fill/vacuum evacuation for at least 2 h, at 100 °C. The water content at this point was found to be below the Karl Fisher limit (<0.02 %wt). In the final step, the system was evacuated to 100 mTorr, closed, and equilibrated at the reaction temperature. 2.4-TDI was melted in a second round-bottom flask and added via syringe to the reaction vessel filled with diol. The vessel content was stirred under nitrogen and after that put to the oven during 48 h at 60 °C. For some networks based on diol M_n =1900 the monool component was also added to obtain modified network precursors (containing the mixture of telechelic diisocynates and monoisocyanates which influences later increasing amount of dangling chains). A small amount of the network precursor was either titrated with dibutylamine (to measure the isocyanate content) or stirred with dry methanol for 24 h, to cap any unreacted isocyanate groups in the samples.

2.3. Network Preparation

In the second step networks were prepared by cyclotrimerisation of telechelic diisocyanate obtained in the first step. The catalyst was added by syringe to the network precursor in the reaction vessel (stirred under nitrogen). After 15 minutes of mixing, the reaction mass was put to the steel molds with teflon plates and transferred to the heated oven. The crosslinked samples were obtained by heating at 80°C during 5 days. The optimal conditions for both steps of network preparation were found earlier (*24*). The remaining unsaturated chain ends (ureacted [NCO] groups) of the samples were blocked in the vapour of methanol.

2.4. Network Characterization

Samples with different concentration of elastically active chains and different content of dangling chains were prepared by varying the diol size or by the content of monool component. The properties of networks were estimated by multiple extractions in xylol, differential scanning calorimetry (DSC), photoelastical measurements and obtained data are given in the Table 1.

2.5. Thermogravimetry Analysis

The thermal degradation of prepared network samples was studied using TGA Perkin Elmer TGS-2 thermal analyser, in air and nitrogen with sample masses of about 3 mg. The heating rate was 10 °C/min, the temperature range was from 30 °C to 500 °C. Additionally, the thermal stability of some extracted and non-extracted samples with diol size M_n=1900, was characterized by simultaneous TG/DSC thermal analyzer SDT Q600 TA Instruments, with the same heating rate and sample mass, but in air and nitrogen gas carriers (100 cm^3/min). TG/MS data were obtained using TA Instruments SDT 2960 setup, coupled with Balzers Termostar GSD 300 T capillary MS in helium carrier, at a heating rate of 10°C/min, with sample masses of about 7 mg in platinum pan.

2.6. Dynamic Mechanical Analysis

Dynamic mechanical properties were examined by Dynamic Mechanical Thermal Analyzer DMTA IV of Rheometric Scientific in the tensile mode, at a frequency of 1 Hz and sample dimensions of about 25 mm (length), 5 mm (width) and 0.120 mm (thickness). The storage modulus, G′, loss modulus, G″, and loss factor, tan delta, were measured using a sinusoidal tensile strain of 0.05% with 0.1% static strain to keep the tension constant. The measurements were carried out in a temperature range from −100 °C to 40 °C, with a heating rate of 2 °C/min.

3. Results and Discussion

3.1. The Structure and Properties of Networks

The urethane prepolymers end-capped with diisocyanate groups and had an exact number of telechelic macromolecular monomers, so-called telechelic macromonomers (figure 1), provide a facile route to design sophisticated polymer molecular structures. The structure of obtained network is shown in the Figure 2. We managed also to prepare model networks with defined pendant chains by partial substitution of the diol with a mono-hydroxy component in the preparation of isocyanate terminated network precursor. It has been found that evaluation of structural parameters for networks obtained by crosslinking reaction of isocyanate groups can be calculated by applying the theory of branching processes (*25*).

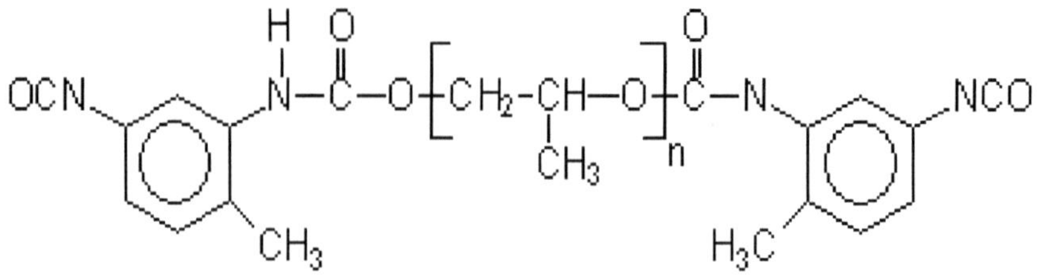

Figure 1. The structure of network precursor based on 2.4-TDI and poly(oxypropylene) diol for r = OH/NCO=0.5.

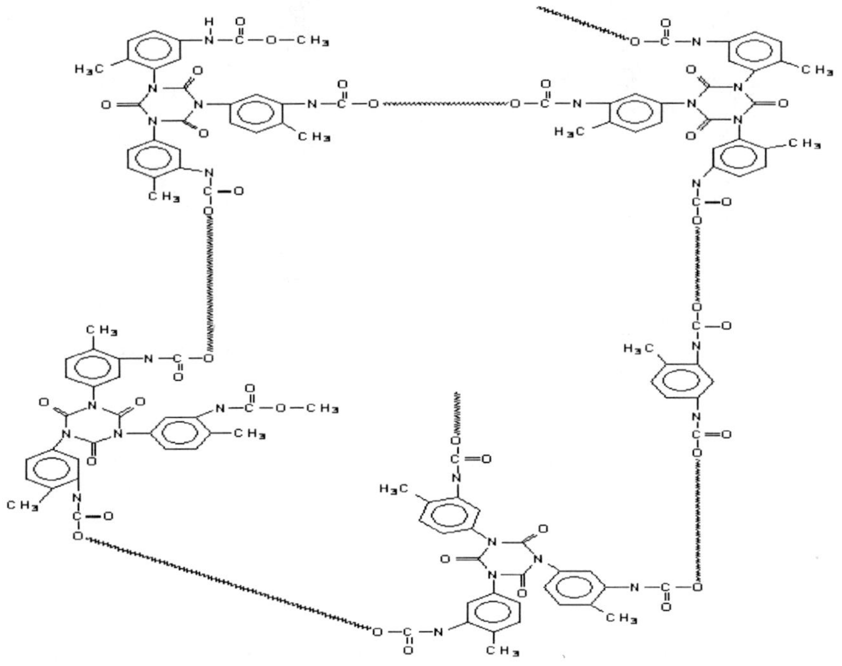

Figure 2. The structure of obtained poly(urethane-isocyanurate) network.

The theory of branching processes with cascade substitution was used for the network structure estimation by taking into account the real functionality of a diol and the existence of the first shell substitution effect on 2.4-TDI. Weight fractions (w), concentrations (v), M_n for elastically active, backbone and dangling chains were estimated (on the basis of experimentally determined gel content) and are summarised in Table 1, together with some experimental data. Cross-linking densities (α_{EANC}) were from 2.8x10^{-4} to 0.1x10^{-4} mol/cm^3 and dangling chains concentrations (ν_{DC}) from 0.9x10^{-4} to 1.6x10^{-4} mol/cm^3. In Figure 3. are shown dependances of M_n and $EANC$ and dangling chain concentration ν_{EANC} and ν_{DC} on gel content during the cyclotrimerisation of telechelic diisocyanates calculated by the theory of branching processes with cascade substitution. It is estimated that concentration of elastically active chains (ν_{EANC}) was decreasing from 5.996 10^{-4}mol/cm^3 to 0.798 10^{-4} mol/cm^3 by increasing diol size. For samples based on modified precursor based on diol M_n = 1900 and monool component the networks with increasing content of dangling chains were obtained. It is determined that with increasing monool content, the dangling chain fraction w_{DC} increased from 0.25435 to 0.5042. As we expected, network based on short diol (PPD 725) at room temperature is in transition region, thus it is elastomeric material only at higher temperature (determined T_g, was –16 °C). The assembly of soluble branching structures of different distribution still exists in non-extracted samples because, for network prepared in the bulk it is not possible to reach the total conversion of [NCO] groups during cyclotrimerisation reaction.

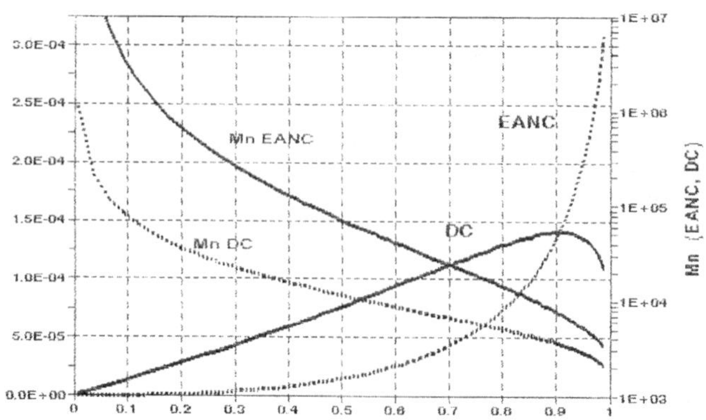

Figure 3. The dependances of ν_{EANC}, ν_{DC}, Mn_{DC} and Mn_{EANC} on gel content w_g during the cyclotrimerisation of telechelic diisocyanates.

Table 1. Experimental data and structural parameters obtained using theory of branching processes based on gel content (w_g) as input data for networks with different content of DEGME, X=[OH$_m$]/([OH$_m$]+[OH$_d$])

	PPD 725	*PPD 1300*	*PPD 2000*	*PPD 2000*	*PPD 2000*	*PPD 2000*	*PPD 4000*
M_n of diol (VPO)	730	1230	1900	1900	1900	1900	3260
X	0	0	0	0.10	0.15	0.30	0
Gel content, w_g	0.983	0.9873	0.9787	0.8964	0.8650	0.7587	0.8951
Final conversion [NCO] α_{trim}	0.885	0.9316	0.9180	0.8552	0.8611	0.9014	0.8664
T_g °C (DSC)	-16	-36	-48	-46	-45	-45	-64
Ge, kPa	1390	948.3	697.0	319.8	269.7	169.7	240.6
Concentration of EANC ν_{EANC}, 10^{-4} mol/cm^3	5.996	4.268	2.783	1.312	1.090	0.855	0.798
Concentration of DC ,ν_{DC}, 10^{-4} mol/cm^3	1.818	1.038	0.8800	1.4041	1.4809	1.542	0.849
weight fraction of DC, w_{DC}	0.228	0.209	0.254	0.439	0.467	0.504	0.444

3.2. Dynamical-Mechanical Analysis

For the estimation of influences of elastically active and dangling chains content on dynamical-mechanical properties (storage modulus, loss modulus and tan δ) of obtained networks, DMA analysis was utilized. When the time scale of the viscoelastic relaxation is comparable to that of the vibration, most of the vibration energy can be dissipated as heat by the polymer. One can use the loss tangent, defined by the ratio G″/G′ of the storage (G′) and loss moduli (G″), as a measure of the dissipation of deformation energy. Figure 4. shows temperature dependences of *tan δ* for networks based on diol *Mn*=1900 and different DEGME content. Temperature dependences of tan δ for samples with different diol size, *Mn*, are presented in the Figure 5. The dependences of *tan δ* is established at 60°C and its value is 3.3 10^6 Pa, while at 20°C it is 9.9 10^6 Pa. Glass transition temperatures are in the range of –48 to –45 °C. Dynamic mechanical properties in the temperature range from –80°C to 40°C, at frequency of 1Hz, show that only for networks with shorter diol (M_n=730) do not achive equlibrium value of storage modulus. Table 2 shows viscoelastic properties of networks determined by dynamic mechanical measurements in the tensile mode at a frequency of 1 Hz. Summarized data for dynamic mechanical properties of networks, determined in the tensile mode, at a frequency of 1 Hz are given in Tables 2 and 3.

Table 2. Viscoelastic properties of networks determined by dynamic mechanical measurements in the tensile mode at a frequency of 1 Hz

M_n diol,	G′ (Pa) (at -80°C)	Tg (°C) from G′	Tg (°C) from G″	G′(Pa) (at +20°C)
M_n = 730 X = 0.0	1.6 10^9	–13	–10	9.9 10^6 (at +60 °C) 3.3 10^6
M_n = 1230 X = 0.0	1.6 10^9	–36	–35	3.0 10^6 -
M_n = 1900 X = 0.0	1.5 10^9	–55	–52	1.9 10^6 -
M_n = 1900 X = 0.3	1.4 10^9	–52	–50	1.6 10^6 -
M_n = 1900 X = 0.5	1.4 10^9	–50	–47	1.3 10^6 -
M_n = 3260 X = 0.0	1.3 10^9	–64	–62	1.05 10^6 -

Table 3. Data for determined glass transitions (T_g) at maximum height of tan δ and estimated network parameters by the theory of branching processes

M_n diol DEGME Content	T_g (°C) at tan δ max	Max tan δ	w_{EANC}	ν_{DC} [10^{-4} mol/cm^3]	w_{DC}	w_{DC}/w_{EANC} (%)
M_n = 730 X = 0.0	+12	0.7	0.9828	1.81	0.205	21
M_n = 1230 X = 0.0	–20	0.9	0.9873	1.04	0.228	23
M_n = 1900 X = 0.0	–43	1.2	0.9787	0.88	0.254	25
M_n = 1900 X = 0.3	–38	1.0	0.7587	1.55	0.504	69
M_n = 1900 X = 0.5	–33	0.8	0.4094	0.93	0.356	89
M_n = 3260 X = 0.0	–55	1.9	0.8988	0.85	0.439	49

It is demonstrated that *tan δ* increases with increasing w_{DC} whose one end is free to move significantly contributes to the dissipation of deformation energy via viscoelastic ralaxation. The temperature insensitive damping originates from a broad relaxation spectrum of the irregular networks comprising the dangling branched chains with various shapes and a wide size distribution. An important key of the energy dissipation in the irregular networks is the fact that one end of each dangling chain is free to move. It is possible to expect that a high irregularity in network structure will yield not only a pronounced damping, but also temperature- and frequency-insensitive damping and elasticity because of a broad distribution of relaxation time. Temperature dependences of *tan δ* for networks with different diol size, *Mn*, are given in Figure 4, and temperature dependences of *tan δ* for networks based on diol *Mn*=1900 and monool are given in Figure 5. In Figure 6 are presented the dependances of damping region calculated from criterium (*tan δ* ≥ 0.5) vs. w_{DC}/w_{EANC} for prepared networks based on different diol size or different content of monool component.

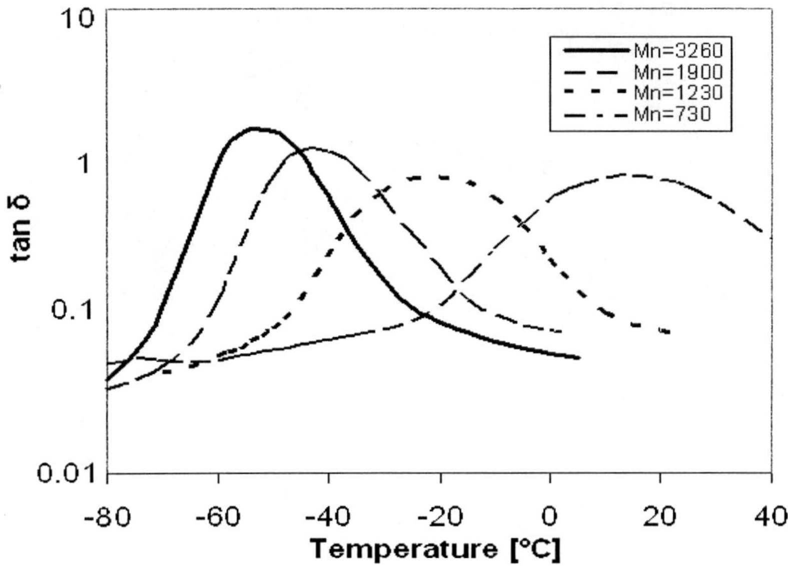

Figure 4. Temperature dependences of tan δ for poly(urethane-isocyanurate) networks with different diol size, Mn.

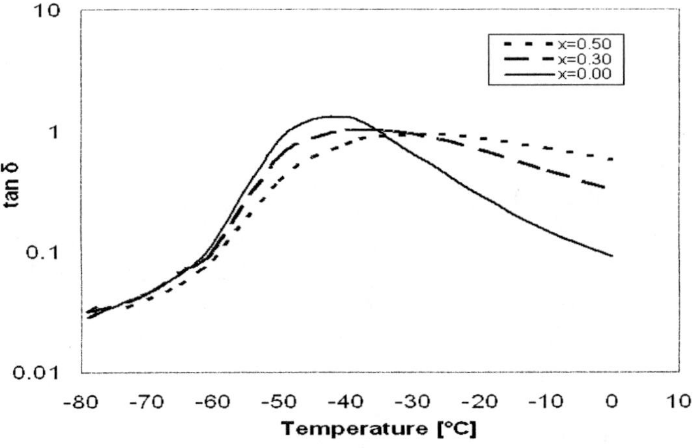

Figure 5. Temperature dependences of tan δ for networks based on 2,4-TDI and diol Mn =1900 and different DEGME content.

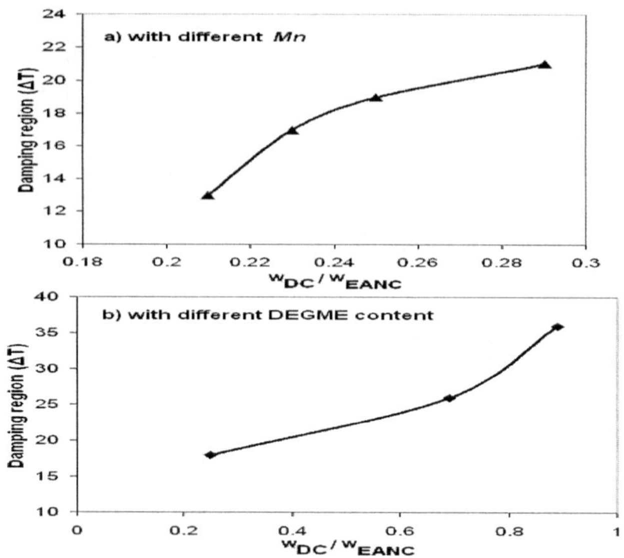

Figure 6. The dependances of damping region on criteria (tan $\delta \geq 0.5$) vs calculated ratio w_{DC}/w_{EANC} for a) networks based on different diol size, Mn, and b) networks based on diol Mn =1900 and different monool content X.

3. 3. Thermal Decomposition

The thermal decomposition of polyurethanes takes place via any of the following routes or more likely through a combination of them: (1) random-chain scission, (2) chain-end scission, i.e., unzipping, and (3) crosslinking (*13*). The main decomposition routes of polyurethane networks are random chain scission and cross-linking. The bond breaking inside the bulk polymeric matrix requires a substantial energy. On the contrary, the enthalpy of breaking the side chain bonds is considerably lower. Therefore, the thermal decomposition of polyurethane starts preferentially via degradation of side chains. The volatile materials trapped inside of the matrix are also easily released at the beginning of the decomposition. It is followed by the depolymerization, resulting in significant weight loss. T. Gupta and B. Adhikari (*26*) have found that with increasing crosslink density of polyurethanes, their thermal stability increases. In general, thermal stability of polyurethane materials depends on the ratio of the urethane group number, formed during telechelic diisocyanate preparation, and the number of isocyanurate rings. However, it was shown that at higher temperatures not necessarily the weakest link in the chains determines the decomposition mechanism but it is often affected by the environment (*27*). The degree of phase separation plays an important role in decomposition of polyurethanes (*28*). By thermogravimetry for poly(urethane-isocyanurate) networks the thermostability of the isocyanureate linkage was found to be significantly higher (150°C) compared to that of the urethane one (*26, 29*). Derivative thermogravimetric curves (DTG) of samples based on different diol size, *Mn*, in air and nitrogen show that the decomposition of the networks highly depends on the diol size and also the gas carrier (*30*). Due to different junctions and groups in the network the decomposition takes place in several overlapping steps. In a nitrogen atmosphere, in the macromolecules with lower *Mn* two overlapping steps are clearly distinguished and the thermal stability of the samples with lower *Mn* is somewhat higher. In oxygen just the opposite trend is observed: the thermal stability of the chains with higher *Mn* is higher. The overlapping decomposition steps are hardly visible. However, the asymmetry of DTG curves of networks with lower *Mn* is still higher. DTG peak maximum is almost independent on diol size, except of that with the lowest *Mn* which has maximum at the highest temperature in nitrogen while the lowest one in air. A typical simultaneous TG/DSC and DTG curve set taken in flowing nitrogen for extracted and non-extracted samples with X=0.25 is presented in Figure 7. The peaks in DTG curves, marked by circles, are the consequence of balance trembling, referring to the network softening in the sample pan. The softening of the sample is accompanied with the beginning of its thermal decomposition. The thermal stability of non-extracted polymer is lower, according to the presence of sol segments, affecting also the softening process of the samples. TG/MS curves of extracted samples with $X = 0.5$ refers to the CO_2 release (peak with 44 and 45 amu) at the first decomposition step from the urethane segment, in the accordance with literature data (*26, 31*) (Figure 8). The next characteristic peak with 88 and 131 amu is related to TDI fragmentation (Figure 9). The thermogravimetric curves of extracted network samples with diol *Mn* =1900 and different monool content in nitrogen and air gas carriers are shown in Figure 10.

The balance tremble here also refers to the softening of the polymer in nitrogen atmosphere, and is accompanied with the splitting of the urethane bond. The decomposition temperature is independent of the network composition as well as the peak temperature of the isocyanureate linkage splitting which appears at about 130°C higher temperatures compared to the urethane bond splitting. In air the decomposition takes place at about 100°C lower temperature with no signs of the softening in DTG curves. The next, overlapping decomposition step is taking place also at lower temperatures. At higher temperatures uncontrolled oxidative processes appear. The kinetics of decomposition at elevated temperature can be extrapolated back to the service temperature for which the lifetime prediction of the material is required. Kinetic data (activation energy and reaction order) of the decomposition reactions were obtained using Freeman-Carroll method (*32*). The data for extracted and non-extracted network samples with *Mn* =1900 diol and different monool content were calculated on the basis of DTG curves, obtained in nitrogen gas carrier for the second stage of the network sample decomposition (degradation of isocyanurate rings). It was assessed that with increase of dangling chain weight fraction (w_{DC} from 0.254 to 0.504) influenced by monool component addition (X from 0.10 to 0.50) the values for the reaction order for the isocyanurate rings decomposition increases from 0.46 to 0.91, and that the values for activation energy for this step are very close. The presence of sol in the non-extracted samples of polyurethenes with monool content (0.20, 0.25 and 0.50) influenced the lower values for kinetic parameters (activation energy and reaction order) compared to values for extracted samples with the same monool content.

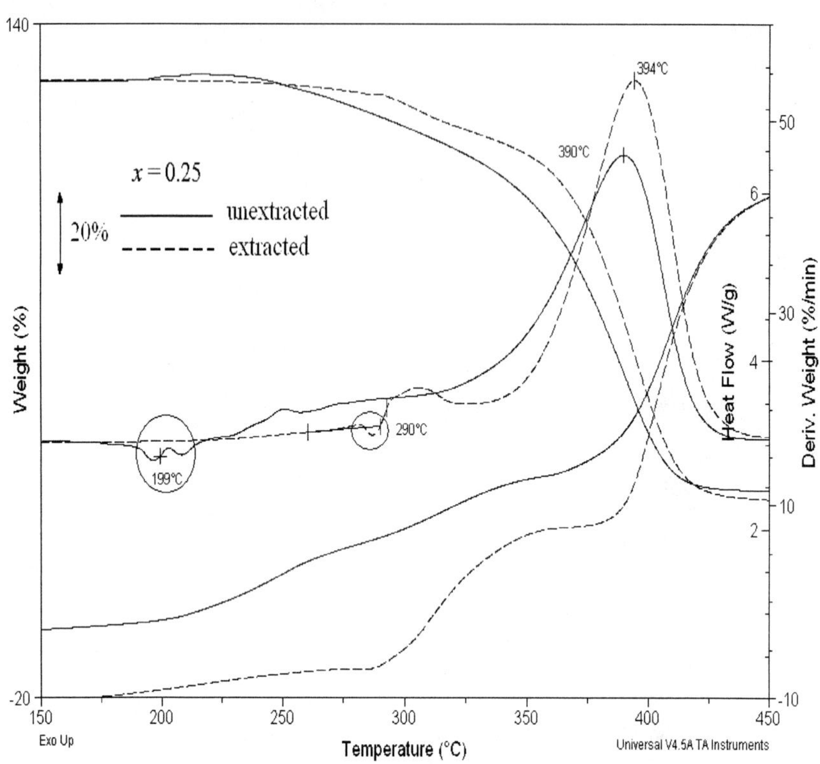

Figure 7. TG/DSC – DTG curves for extracted and non-extracted network with diol Mn =1900 and X=0.25 in nitrogen gas carrier.

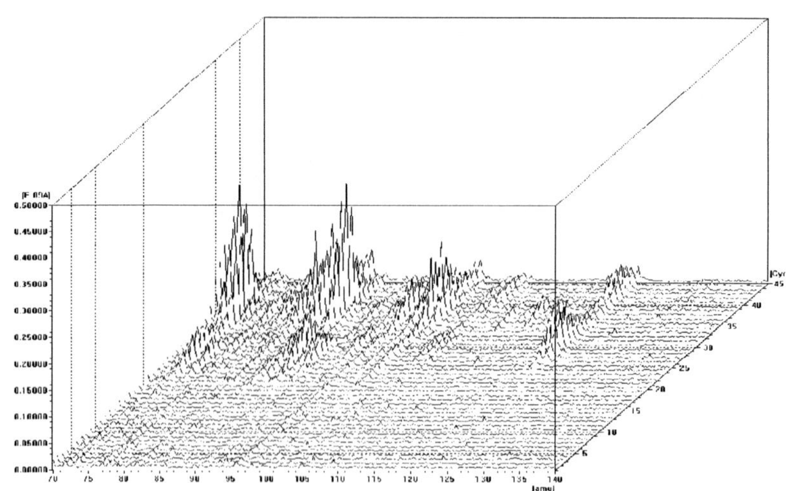

Figure 8. 3D TG/MS analysis of extracted network sample with diol Mn =1900 and X=0.5, for 35 – 70 amu range.

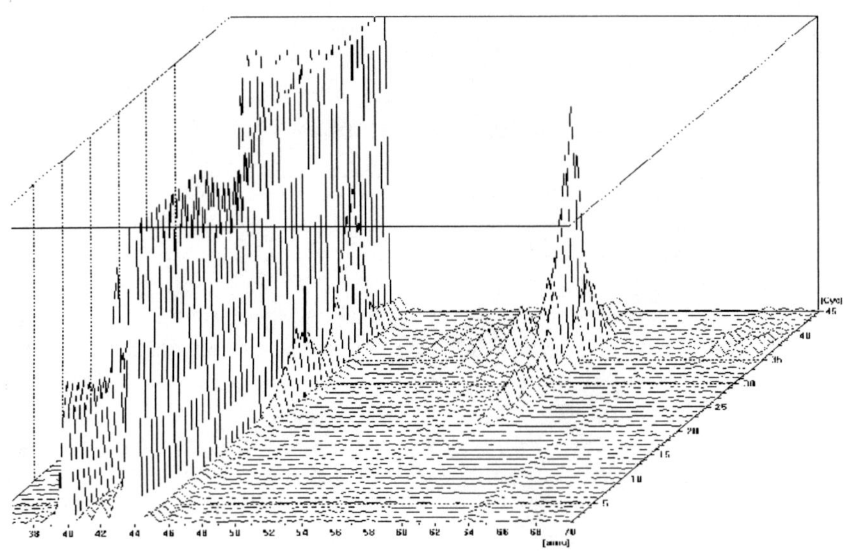

Figure 9. 3D TG/MS analysis of poly(urethane-isocyanurate) network with diol Mn =1900 and X=0.5, for 70 – 140 amu range.

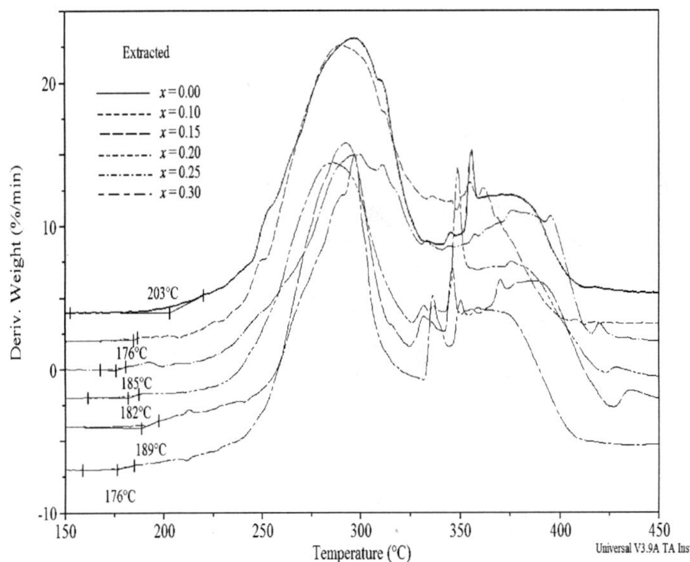

Figure 10. DTG curves in air of extracted network samples with diol Mn=1900 for different monool content, X.

4. Conclusions

In this work, attention has been focused on preparation polyurethanes which are suitable for the applications that require heat resistance of materials. For this purpose, the network with isocyanurate(hexahydro-1,3,5-triazin-2,4,6-trion) rings were prepared by cyclotrimerisation of telechelic diisocyanates. The network topology was calculated using the stochastic theory of branching processes. It is estimated that concentration of elastically active chains decreased from 5.996×10^{-4} mol/cm^{-3} to 0.798×10^{-4} mol/cm^{-3} by increasing the diol size *Mn* from 730 to 3260. For networks based on precursor prepared by adding monool component, the increasing content of dangling chains was obtained. Networks with increasing content of pendant chains have potentially good energy absorption properties as it was assessed from calculated damping criterium $\tan\delta \geq 0.5$ from dynamic mechanical measurements. Using the thermogravimetry, it was estimated that isocyanurate rings enhanced thermal stability. It was determined that the decomposition of networks processed at least in two steps. The first step is decomposition of urethane bond and the second step is decomposition of isocyanurate ring. Despite the monool content, the onsets of degradation for all networks with diol size *Mn*=1900 appeared to start at the same temperature. The kinetics of decomposition at elevated temperature can be extrapolated back to the service temperature for which the lifetime prediction of the material is required.

References

1. Dušek, K. Networks from Telechelic Polymers: Theory and Application to Polyurethane. In *Telechelic Polymers: Synthesis and Applications*; Goethals, E. J., Ed.; CRS Press: Boca Raton, FL, 1989; pp 289–360.
2. Ilavsky, M.; Somvarsky, J.; Bouchal, K.; Dusek, K. Structure, equilibrium and viscoelastic mechanical behaviour of polyurethane networks based on triisocyanate and poly(oxypropylene)diols. *Polym. Gels Networks* **1993**, *1*, 159–184.
3. Urayama, K.; Miki, T.; Takigawa, T.; Kobjiya, S. Damping elastomer based on model irregular networks of end-linked poly(dimethylsiloxane). *Chem. Mater.* **2004**, *16*, 173–178.
4. Šomvarsky, J.; Nijenhuis, K.; Ilavsky, M. Polyfunctional cross-linking of existing polymer chains. *Macromolecules* **2000**, *33*, 3659–3670.
5. Budinski-Simendić, J.; Ilavsky, M.; Šomvarsky, J.; Špirkova, M.; Korugic-Karasz, Lj.; Radičević, R.; Dikić, T.; Dušek, K. Rubber elasticity study of polyurethane networks prepared by endlinking of telechelic diisocyanates. *Mater. Sci. Forum* **2007**, *555*, 491–496.
6. Dušek, K.; Dušková-Smrčková, M. Polymer networks from precursors of defined architecture: Activation of preexisting branch points. *Macromolecule* **2003**, *36*, 2915–2925.
7. Dušek, K.; Šomvarsky, J. Modelling of ring-free crosslinking chain (co)polymerization. *Polym. Int.* **1997**, *44*, 225–236.

8. Dušek, K.; Dušková-Smrčková, M.; Huybrechts, J. Control of performance of polymer nanostructured polymer precursors by differences in reactivity of fuctional groups. *J. Nanostruct. Polym. Nanocomp.* **2005**, *1*, 45–53.
9. Heijkantsa, R. G. J. C.; Calcka, R. V. V.; Tienenb, T. G. V.; Groota, J. H. D.; Bumab, P.; J. Penningsa, J. A.; Veth, R. P. H.; Schoutena, A. J. Uncatalyzed synthesis, thermal and mechanical properties of polyurethanes based on poly(ε-caprolactone) and 1,4-butane diisocyanate with uniform hard segment. *Biomaterials* **2005**, *26*, 4219–4228.
10. Chattopadhyay, D. K.; Webster, D. C. Thermal stability and flame retardancy of polyurethanes. *Prog. Polym. Sci.* **2009**, *34*, 1068–1133.
11. Król, P.; Pilch-Pitera, B. Mechanical properties of crosslinked polyurethane elastomers based on well-defined prepolymers. *J. Appl. Polym. Sci.* **2008**, *107*, 1439–1448.
12. Tatai, L.; Moore, T.; Adhikari, R.; Malherbe, F.; Jayasekara, R.; Griffiths, I.; Gunatillake, P. A. Thermoplastic biodegradable polyurethanes: The effect of chain extender structure on properties and in-vitro degradation. *Biomaterials* **2007**, *28*, 5407–5417.
13. Laércio, G. L.; Yoshio, K. Thermal degradation of biomedical polyurethanes: A kinetic study using high-resolution thermogravimetry. *J. Appl. Polym. Sci.* **2001**, *79*, 910–919.
14. Correa, R.; Nunes, R.; Lourenco, V. Investigation of the Degradation of thermoplasti urethane reinforced with short fibers. *Polym. Degrad. Stab.* **1996**, *52*, 245–251.
15. Kim, H.; Lee, T.; Huh, J.; Lee, D. Preparation and properties of segmented thermoplastic polyurethane elastomers with two different soft segments. *J. Appl. Polym. Sci.* **1999**, *73*, 345–352.
16. Desai, S.; Thakore, I. M.; Sarawade, B. D.; Devia, S. Efect of polyols and diisocyanates on thermo-mechanical and morphological properties of polyurethanes. *Eur. Polym. J.* **2000**, *36*, 711–725.
17. Takeichi, T.; Guo, Y.; Agag, T. Synthesis and characterization of poly(urethane-benzoxazine) films as novel type of polyurethane/phenolic resin composites. *J. Polym. Sci.* **2000**, *38*, 4165–4176.
18. Sergei, V. L.; Edward, D. W. Thermal decomposition, combustion and fire-retardancy of polyurethanes: A review of the recent literature. *Polym. Int.* **2004**, *53*, 1585–1610.
19. Benli, S.; Yilmazer, U.; Pekel, F.; Ozkar, S. Effect of fillers on thermal and mechanical properties of polyurethane elastomer. *J. Appl. Polym. Sci.* **1998**, *68*, 1057–1065.
20. Sudip, R.; Allan, J. E. Advances in polymer-filler composites: Macro to nano. *Mater. Manuf. Processes* **2007**, *22*, 741–749.
21. Semsarzadeh, M.; Navarchian, A. Morshedian, reactive extrusion of poly(urethane-isocyanurate). *Adv. Polym. Technol.* **2004**, *23*, 239–255.
22. Ludwick, A.; Aglan, H.; Abdalla, M.; Calhoun, M. Degradation behavior of an ultraviolet and hygrothermally aged polyurethane elastomer: Fourier transform infrared and differential scanning calorimetry studies. *J. Appl. Polym. Sci.* **2008**, *110*, 712–718.

23. Budinski-Simendić, J.; Ilavsky, M.; Špirkova, M.; Šomvarsky, J.; Dušek, K. Modelling of the structure and properties of poly(urethane-isocyanurate) elastomeric materials. *Mater. Sci. Forum* **1998**, *282*, 295–302.
24. Špirkova, M.; Budinski-Simendić, J.; Ilavsky, M.; Špaček, P.; Dušek, K. Formation of poly(urethane-isocyanurate) networks from poly(oxypropylene) diols and diisocyanate. *Polym. Bull.* **1993**, *31*, 83–88.
25. Budinski-Simendić, J.; Špirkova, M.; Dušek, K; Dikić, T.; Radičević, R.; Prendzov, S.; Krakovsky, I.; Ilavsky, M. The properties of dangling chain networks prepared by cyclotrimerization of telechelic diisocyanates. *Mater. Sci. Forum* **2006**, *518*, 399–404.
26. Gupta, T.; Adhikari, B. Thermal degradation and stability of HTPV-based polyurethanes and polyurethaneureas. *Thermochim. Acta* **2003**, *402*, 169–181.
27. Pielichowski, K.; Słotwińska, D.; Dziwiński, E. Segmented MDI/HMDI-based polyurethanes with lowered flammability. *J. App. Polym. Sci.* **2004**, *91*, 3214–3224.
28. Wang, T.; Hsieh, T. Effect of polyol structure and molecular weight on the thermal stability of segmented poly(urethaneureas). *Polym. Degrad. Stab.* **1997**, *55*, 95–102.
29. Pavličević, J; Budinski-Simendić, J.; Radičević, R.; Katsikas, L.; Popović, I.; Mészáros Szécsényi, K.; Špirkova, M. Preparation and themal stability of elastomers based on irregular poly(urethane-isocyanurate) networks. *Mater. Manuf. Processes* **2009**, *24*, 1217–1223.
30. Montaudo, G.; Puglisi, C.; Scamporrino, E.; Vitalinit, D. Mechanism of thermal degradation of polyurethanes: Effect of ammonium polyphosphate. *Macromolecules* **1984**, *17*, 1605–1614.
31. Semsarzadeh, M.; Navarchian, A. Effects of NCO/OH ratio and catalyst concentration on structure, thermal stability, and crosslink density of poly(urethane-isocyanurate). *J. Appl. Polym. Sci.* **2003**, *90*, 963–972.
32. Freeman, E.; Carrol, B. The applications of thermoanalitical techniques to reaction kinetics: The thermogravimetric evaluation of the kinetics of the decomposition of calcium oxalate monohydrate. *J. Phys. Chem.* **1958**, *62*, 394–397.

Chapter 12

The Effect of Filler Particles on the Properties of Elastomeric Materials Based on Different Network Precursors†

Vojislav Jovanović,[1] Jaroslava Budinski-Simendić,[*,2] Jelena Milić,[2] Ayse Aroguz,[3] Ivan Ristić,[2] Slobodan Prendzov,[4] and Ljiljana Korugic-Karasz[5]

[1]Faculty of Science, Kosovska Mitrovica, Serbia
[2]University of Novi Sad , Faculty of Technology, Serbia
[3]University of Istanbul, Engineering Faculty, Turkey
[4]Faculty of Technology and Metallurgy, Skopje, Macedonia
[5]Department of Polymer Science and Engineering, University of Massachusetts, Amherst, Massachusetts, USA
[*]jarka@uns.ac.rs

†Dedicated to Professor F. E. Karasz on the occasion of his 75th birthday

The incorporation of nanoparticles to rubber compound with subsequent conversion to a network by crosslinking leads to a materials of great complexity. As compared to micron size filler particles the nano size filler particles are able to occupy substantially greater number of sites in the polymer matrix. Selecting proper network precursors for a specific application requires an evaluation of the environment in which it is anticipated to operate and the performance required. Compared to carbon black, silica has two advantages for fundamental investigations: it has a simpler geometry and it can be functionalized in different ways to control its degree of interaction with the polymer (i.e.), compatibility to enhance dispersion and/or mechanical coupling by covalent or non-bonded interactions. In this contribution, studies of elastomers based on ethylene-propylene diene monomer (EPDM) rubber and acrylonitrile butadiene rubber (NBR) have been performed. To enhance the quality of elastomers, an attempt was made to prepare technologically feasible blends comprising non-polar EPDM rubber and NBR with high performance. The properties of elastomeric composites were determined before and after thermal aging in an air circulating oven. The elastomers based on EPDM/NBR (80:20) were reinforced either with high abrasion furnace carbon black (46 nm) or with precipitated silica (22 nm) or its combination. Comparative values of tensile strength before and after weathering were analyzed. It was estimated considerable performance enhancement of rubber blends by reinforcing with dual active fillers.

Introduction

With the exception of network formation, no process in the elastomers manufacturing is more important as the reinforcement by particulate fillers (especially carbon black and silica). Fillers are generally incorporated into elastomeric materials to modify the mechanical and electrical properties, while keeping the cost of the end product low. Success is dictated by selecting a network preursor having good thermal stability that meets the environmental requirement for which the application is intended. The elastomers used for wire and cable insulation and jacketing or sealant manufacture are subjected to an accelerated degradation under the action of environmental factors and/or electrical field. In immiscible rubber blends, the situation at the polymer interface is critical as a high-interfacial tension and a poor adhesion between the phases are observed. Compatibilization was found to improve the properties of immiscible blends by providing stable morphology, and good interfacial adhesion (*1*). Elastomers based on more network precursor have become technologically important materials for their diverse applications. Their physical and chemical properties recommend them as engineering materials for electric insulators, chemical industry, automotive and aircraft production, and many other areas. To enhance the physical properties of blends, it is important to understand the atomic-scale microstructure of the related polymers. Nitrile butadiene rubber (NBR) is a copolymer of butadiene and acrylonitrile and it comes under "special purpose" synthetic rubbers. The electrical, mechanical, and nanoscale free-volume properties of NBR and EPDM rubber reinforced by bentonite and kaolin were studied (*2*). Thermal properties of ethylene-propylene elastomers and acrylonitrile-butadiene rubber blends were studied in ref. (*3*). Determination of the compatibility of NBR/EPDM blends by an ultrasonic technique and modulated DSC were investigated by Mathur and co-workers (*4*). Filler distribution affects the properties of the blend, as it is controlled by the molecular weight of the polymer and filler dispersion in each phase and chemical interaction between polymer and filler (*5*). Elastomers are insulators because the atoms in the rubber macromolecule chain are covalently bonded. In the covalent bonded molecules of saturated carbon compounds there is no scope of delocalization of the valence electrons, consequently, the electron carrier path is not available. One of the important methods to form carrier path in an insulating rubber matrix is the incorporation of conductive additives like carbon black, carbon fiber, metal or metal oxide. One of the used fillers is carbon black; it is selected because it is easily processed and also produces a reinforcing effect on the materials. These conductive composite materials have been of low cost, high flexibility, and good mechanical properties. These are widely used for rubber contact switches, floor heating, electromagnetic interference shielding and various other electronic and electrical applications. The curing process is the final step in the elastomers manufacturing whereby a rubber goods are formed to the desired shape. In press, heat is transferred to the compounds from the surfaces, which are maintained at high temperatures, inducing the network formation i.e. obtaining the strong elastic material. The study of structure-property relationships for elastomeric nano-composites has acquired significance due to the broad range of applications as advanced materials. A detailed understanding of the general mechanism of rubber reinforcement is one key to develop new systems providing the performance enhancement needed (*6*, *7*). The low modulus of elastomers combined with their high reversibility of deformation allows their use in a wide range of industrial applications such as tires, hoses, belts, drug delivery devices and products such as rubber tubes, sealants, biomedical and marine devices. Developing elastomeric products includes efforts of materials engineering with the aim to improve the properties of a specific rubbers for a certain application (*8*). Elastomeric composites show benefit both from the elasticity of the rubbery component and from the stiffness of the reinforcing (*9*). Although the reinforcement of rubber by active fillers is a well-recognized phenomenon the term 'reinforcement' is not well defined. Briefly it can be stated that it means the pronounced increase in tensile strength, tear resistance, abrasion resistance and modulus far beyond the values expected on the basis of the Einstein-Guth and Gold theory, taking into account the effects caused by colloidal spherical particles (hydrodynamic effect) and occlusion of rubber. The reinforcement by finely divided fillers, particularly carbon black and silica, is fundamental to the rubber industry (*10*, *11*). The polymer network contribution depends on the cross-linking density of the matrix and the nature of the network precursors. In the hydrodynamic model is nothing else than the effect of strain amplification, resulting from the fact that the filler is the rigid phase, which cannot be deformed. As a consequence, the intrinsic strain of the polymer matrix is higher than the external strain yielding a strain-independent contribution to the modulus. The effect of the structure is attributed to the 'in-rubber structure', which can be understood as a combination of the structure of the filler in the in-rubber state and the extent of filler–polymer interaction. The in-rubber structure is the measure for the occluded rubber, which is shielded from deformation and therefore increases the effective filler content leading also to a strain-independent contribution to the modulus. The filler–polymer interaction can be attributed to physical (van der Waals) as well as to chemical linkages or a mixture of both. In the case of the silica–silane system this interaction is formed by chemical linkages. The stress softening at small amplitudes is attributed to the breakdown of the inter-aggregate association respectively to the breakdown of the filler network. This stress softening at small deformations, called Payne-effect (*12*), plays an important role in the understanding of reinforcement mechanism of filled rubber samples. Silica alone is building up a strong filler network and shows a low interaction with the polymer. But, if a two-functional silane is used chemical linkages between the silica and polymeric matrix were established. The bondings are of chemical nature, and therefore macromolecular segments cannot undergo surface displacements by molecular slippages. In the case of carbon black linkages between the surface and polymer chains are formed only by physical adsorption. In the past as well as in the present molecular slippages along the carbon black surface as well as adsorption and desorption mechanisms are still under discussion. In-rubber structure is interpreted as a combination of the effects caused by the real carbon black structure in the in-rubber state and the filler–polymer interaction. Medalia has proposed (*13*) that in a carbon black–rubber system the macromolecules which fills the void space within each aggregate was occluded and shielded from deformation and thus acts as part of the filler rather than as part of the deformable matrix. Due to this phenomenon, the effective volume of the filler is increased considerably and should of course enhance the high strain modulus significantly. As a consequence, the term in-rubber structure explains immediately the term of reinforcement, defined as modulus 300% divided by modulus 100% in a simple stress–strain measurement. Reinforcement is a measure for the strength of a filler to keep its in-rubber structure respectively to restrain the occluded rubber at increasing deformation. From a consideration of the effects of particle size as such, it appears that reinforcement, in the sense of tensile enhancement, will occur with any very finely divided filler. Physical factors prevent escape of the macromolecules from the filler surface (vacuole formation) but allow stress delocalization through interfacial slippage. Occasional stronger bonds may be introduced advantageously to facilitate dispersion, reduce particle/particle interactions, and optimize practical properties relating to resilience and durability. Several lines of evidence suggest that only a minor amount of strong bonding is necessary or desirable, such that polymer/filler slippage can occur, under stress, over most of the interfacial area. Elastomeric composites show benefit both from the elasticity of the rubbery component and from the stiffness of the reinforcing filler. Quite a large variety of powdered minerals can be used for compound preparation but not all have reinforcing capabilities, and essentially two classes of powdered minerals have been found to offer significant reinforcing effects: carbon blacks and high-structured silica. Reinforcement concerns finished elastomeric parts; but it is quite remarkable that flow properties of filled compounds begin to significantly differ from those of unfilled compounds when the filler has reinforcing capabilities. In addition to usual hydrodynamics effects, reinforcing fillers impart indeed other modifications in flow properties whose origin is assigned to strong interactions arising between the rubber macromolecules and the filler particles (*14*, *15*). Recognition of the role of the main factors influencing interfacial adhesion and proper surface modification may lead to significant progress in many fields of research, as well as in related technologies (*16*, *17*). The introduction of even more unconventional fillers, such as carbon nanotubes, metallic, semi-conducting or magnetic nano-particles and clay platelets promises to open a vast array of completely new applications (*18*). Carbon black and silica have been used as the main reinforcing fillers that increase the usefulness of elastomeric materials. As each filler possesses its own advantages, the use of both fillers should enhance the mechanical properties of elastomeric nanocomposites. However, the ratio of active fillers giving rise to the optimum properties needs to be estimated. The progress in tire industry was paralleled by a substantial progress in carbon black technology and production facilities as well. With the ever increasing tire technology development costs and complexity, it became mandatory, especially in the second part of the last century, to replace the trial and error development approach with a more rational way of obtaining competitive products. It then became obvious that a better and fundamental understanding of the materials used was required. Apart from its importance for understanding the reinforcement mechanism, filler distribution is also an indicator of the quality of rubber processing. Fillers with different grades of surface area (elementary particle size) and structure (aggregate shape, branching) will differ in processibility, as indicated by the degree of their incorporation into the matrix, the breakdown of their agglomerates into aggregates and the random distribution of those aggregates. Diverse steps of rubber processing influence filler dispersion in different ways. Elastomers based on Ethylene–propylene–diene monomer (EPDM) rubber exhibits excellent resistance to weather, ozone, acids, and alkalies while accommodating high volume fractions of filler and liquid plasticizers and retaining desirable physical and mechanical properties. These characteristics have allowed elastomers based on this network precursor to be employed as a membrane material for separation operations, which primarily include separation of organics from aqueous streams. Especially elastomers based on 5-ethylidene-2-norbornene (ENB) are very sensitive to oxidation, but is more stable than conventional elastomers (butadiene, isoprene rubber). The addition of non-reinforcing fillers to the material based on this network precursor can reduce sealant friction without significant detriment to the modulus and the tensile strength. Quite different elastomeric materials based on acrylonitrile butadiene rubber (NBR) cannot be used in specific applications requiring high heat and ozone resistance (*19*). The poor ozone resistance and heat ageing properties of this random copolymer of acrylonitrile and butadiene are believed to be the result of unsaturation in the backbone of the macromolecules which permits scission of the chains to occur under certain adverse conditions. The elastomers based on this network precursor are used in automobile industry because of its resistance to fuels, a variety of oils over a wide range of temperatures (*20*) in products like oil seals, water pump seals, blow out preventors, fuel lines, hoses, fuel pump diaphragms, etc. because of its high oil, solvent and fuel resistance and gas impermeability. In tis elastomers there is not present so called self-reinforcing effect, as NBR do not cristallizeas ne. Thus the unfilled vulcanizates have very low tensile strength but when used in combination with reinforcing fillers, materials with excellent mechanical properties can be obtained (*21*).

Although the original purpose of filler was the cost of the compounds, prime importance is now attributed to selectively active fillers and its quantity that produce specific improvements in physical properties (*22–24*). Oxygen containing chemical groups on the carbon black surface can react with the carboxyl groups of some rubbers during high temperature molding of the rubber-filler mixtures (*25*). Carbon black, generally supplied as pellets has to be dispersed during mixing into smaller entities (such as agglomerates and aggregates). Large particle-particle interactions result in inhomogeneous dispersion and distribution of the filler, processing problems, poor appearance and inferior properties. Highly sophisticated process technology was the key to enhance the surface activity of these carbon black grades. This progress in process technology resulting in the structuring of carbon black particles was used to generate special carbon blacks for different applications (*26*). The effect of added carbon black is directly related to the properties of the inter phase and due to the specific interactions between rubber macromolecules and active filler surfaces. It is well known that the properties mainly depend on the dispersion condition of filler particles and their principal relevant properties: particle size, surface area, aggregate structure, surface activity and on rubber-filler interactions (*27*, *28*). It is obvious that to obtain as many interactions as possible the polymer active segments must have access to the majority of the carbon black active sites. Since the mono-unit of carbon black is the aggregate the most possible isolated aggregate must be made available to be in contact with the polymer chains and their unsaturation domains. In the Figure 1 is given the structure of filler agglomerates and location for trapped rubber macromolecules. Networking of fillers depends on its attractive potential and the distance between aggregates. Since, with regard to surface energy, the attractive potential is almost the same for all furnace blacks in hydrocarbon elastomers, the distance between aggregates becomes the controlling factor for compounds processability. Carbon black for the tire industry is today exclusively obtained using the furnace oil process. The apparent density of those particles is extremely low and it would be very difficult to transport such material in an economical and reasonable manner. Compared to carbon black, silica has two advantages for fundamental investigations: it has a simpler geometry and it can be functionalized in different ways to control its degree of interaction with the polymer (i.e.) compatibility to enhance dispersion and/or mechanical coupling by covalent or non-bonded interactions. The specific surface area concept, in active filler physics is very well defined. It represents the total surface area of a unit mass of the material and represents theoretically the total surface in contact with its environment. If it is easy to define this characteristic property, it is much more difficult to measure it; especially in non perfect surfaces, which are characterized by a high surface roughness and/or a heterogeneous energetic sites distribution. Interfacial interactions and interphases play a key role in all multicomponent materials irrespectively of the number and type of their components or their actual structure. They are equally important in particulate filled polymer, fiber reinforced advanced composites, nanocomposites or elastomer based on rubber blends (*29*). The reevaluation of fumed silica becomes of interest since this product is now available in a granulated version. A special aggregate morphology in combination with a narrow particle size distribution and a reduced number of silanol groups on the surface are the main differences to precipitated silicas. However, the ratio of active fillers giving rise to the optimum properties needs to be estimated.

An important characteristic of any elastomer for application purposes concerns its thermal stability, especially when considering the potential end uses in varying fields. The thermal stability of a material is defined by the specific temperature or temperature-time limit within which the material can be used

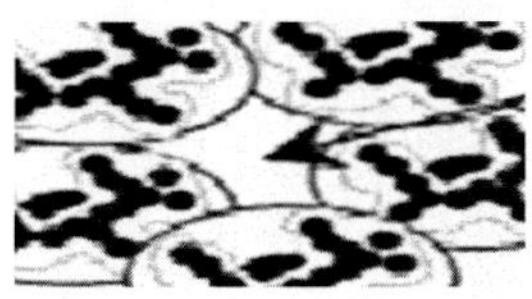

Figure 1. The structure of active filler agglomerates and marked location for trapped rubber macromolecules

without excessive loss of properties. With respect to commercial applications, the investigation of thermal decomposition processes has two important aspects. The first concerns are the stabilization of a polymer to obtain novel materials with a desired level of thermal stability/flame resistance that will be able to fulfill the demands of contemporary materials engineering. To fulfill demanded performance properties as weather resistance and physical properties rubber products can be produced in combination of dense and cellular materials. The choice of the basic polymer in combination with a proper selection of fillers, plasticizers, and curatives are the key criteria to produce the advanced sealing materials for different applications and operating environments. Dynamics or pressure within a system may compel the use of more complex sealing profiles or composite seal assemblies, consisting of multiple materials that, in combination, can offer the required properties. Many differing grades of material, with significantly varying properties, may belong to a single material designation. Elastomers based on more network precursors are widely used in rubber products for a variety of reasons which include improved physical properties, improved service life, easier processing, and reduced product cost (*16*, *30*, *31*). Ghosh studied the properties of EPDM/NBR elastomers in presence of a additive bis(diisopropyl)thiophosphoryldisulphide (*32*). It was estimated that polarity difference between the constituting rubbers of the blend could be narrowed through the introduction of rubber bound by intermediates in the non-polar EPDM. Thermal analysis by modulated differential scanning calorimetry indicated that the segmental immiscibility between NBR and EPDM, and the addition of the compatibilizer resulted in characteristic peak shifts of the individual network precursors to form a broad plateau (*33*). In our earlier investigation we studied the effects of filler and blowing agent content on the morphology of microcellular EPDM elastomers cured in salt bath or hot air (*34*). The research in this applicative work was focused on elastomers based on ethylene-propylene-ethylidenenorbornene rubber intended for sealant production. In the first step materials based on EPDM filled with active filler carbon black and inactive filler chalk were studied. In the second step elastomers based on two network precursors (EPDM and NBR) were studied. This type of materials were reinforced either with one type of filler (high abrasion furnace carbon black or precipitated silica or its combination).

Experimental

Materials

Network precursors: (a) Poly(ethylene-co-propylene-co-2-ethylidene-5-norbornene), containing 11 % ethylidene norbornene, (Vistalon 9500, ExxonMobil Chemical, France), Mooney viscosity ML (1+4) at (125°C) 69.6 Mu (Figure 2). (b) Acrylonitrile-butadiene rubber, Kraynac 34-50, "Polysar" Canada) with acrylonitrile content 33,2%: Moony viscosity at 125°C, 50; specific gravity 1.17 gcm^{-3}; ash content 0,5 % (Figure 3). Active Fillers: (1) High abrasion furnace carbon black, N-330 (Deggusa, Germany): primary particle size 46 nm; specific gravity 1.78-1.82 gcm^{-3}. CTAB surface area 83±6 m^2g^{-1}. (2) Precipitated silica PERKASIL A KS 408 (Deggusa, Germany) a particle size of 22 nm and specific gravity 1.78-1.82 gcm^{-3}. BET surface area 180 m^2g^{-1}. For EPDM elastomers as inactive filler chalk was used hydrophobic, ground natural calcium carbonate, Kredafil 150 Extra "S", (Industrochem Pula, Croatia). Relative density at 20 °C is 2.7, particle size distribution 1-30 micron. Accelerators: N-cyclohexyl-2-benzothiazole sulphenamide (CBS): pale grey; non hygroscopic powder; melting point 95-100°C, specific gravity 1.27-1.31 gcm^{-3}. Tetramethylthiuram disulfide (TMTD) white powder, melting point 140°C, specific gravity 1.33-1.40 gcm^{-3}. Antioxidants: N-isopropyl-N-phenyl-p-phenylene diamine (Vulkanox 4010 NA), specific gravity 1.14-1.18 gcm^{-3}. Curing agent: sulphur: pale yellow powder purity 99.9 %; melting point 112°C; specific gravity 2.04-2.06 gcm^{-3}. Activators: zinc oxide: fine powder; purity 99%, specific gravity 5.6 gcm^{-3}. Stearic acid: melting point 67-69°C; specific gravity 0.838 gcm^{-3}.

Curing Assessment

The curing properties of different rubber compounds were assessed using a Monsanto Oscillating Disc Rheometer R-100. The curing time was adjusted by considering the time needed for the 90% crosslinking at T=160°C. The cure characteristics: ML (minimum torque), MH (maximum torque), t_{c90} (optimum cure time), t_{s2} (scorch time) were registered and cure rate index (CRI) was calculated according equation 1.

$$CRI=\frac{100}{t_{c90}-t_{s2}} \qquad \text{......................} (1)$$

Sample Preparation

To perform a systematic study, elastomeric materials were prepared and cross-linked under same experimental conditions. The rubbers were separately masticated for about 1 min each, keeping a tight nip gap (0.8 mm), and subsequently blended during 3 minutes. The mixing of the rubber macromolecules and filler particles was performed on a two-roll mill at 50 °C, according to the compounding procedure. The compounds were prepared using a laboratory-size

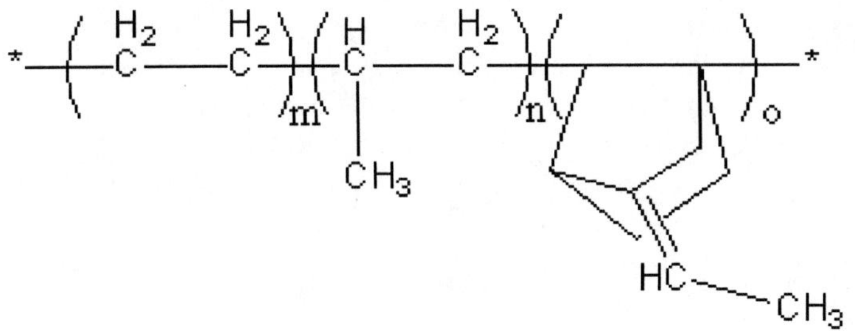

Figure 2. *The structure of ethylene-propylene-ethylidenenorbornene rubber*

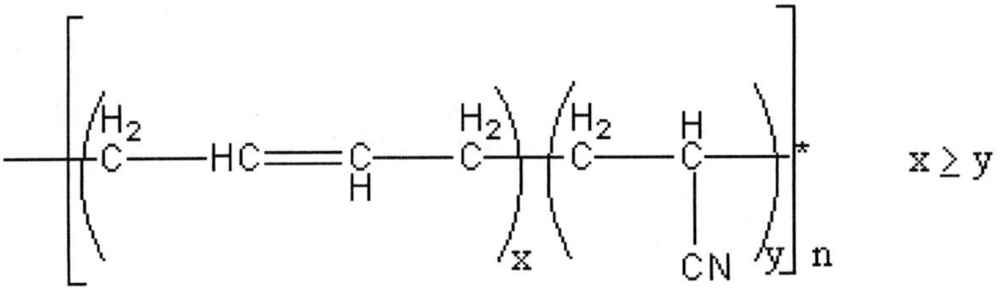

Figure 3. *The structure of acrylonitrile-butadiene rubber*

two-roll mill maintained at 40±5°C. The roller speed ratio was n1/n2=28/22. The mixing time was 20 minutes. The sheeted compound was conditioned at 23±2 °C during 24 h prior crosslinking. The cross-linking of the rubber compounds was carried out in an electrically heated hydraulic press using a mould at a temperature of 160°C for 10 min under a pressure of 10 MPa. Samples in a sheet form were obtained.

Stress/Strain Measurements

For mechanical properties determination all test specimens were compression molded at 160°C during the respective optimum cure time (t_{c90}). The dumbbell samples were cut from a 2 mm thick molded sheet. The mechanical properties were obtained from stress/strain measurements by Tensile tester according ASTM D412-98a. The cross head speed for the tensile test was set at 500 mm/min.

Hardness Measurements

Hardness was measured using an indentation hardness tester. The given results are the mean value of three separate specimens. The error in these measurements is + 5%. Samples with flat surface were cut for hardness test. The measurement was done using Durometer Model 306L Type A.

Differential Scanning Calorimetry

For determination of temperature transitions was used differential scanning calorimeter, DSC Q100, V9.7 Build 291 (TA Instruments), modul DSC Standard

Cell FC. DSC calibration included: T-zero calibration (with empty cell (baseline), and the other with sapphire discs on both sample and reference position; enthalpy (Cell) constant calibration and temperature calibration are based on a run in which indium (metal standard) is heated through its melting transition. Step change (Tg) has been determinate between onset and end, with step midpoint at half height. Heat effect of the phase transition, i.e. normalized enthalpy has been determined by linear peak integration.

Dynamic Mechanical Analysis

The dynamic mechanical spectra of various samples based on EPDM were obtained using a dynamic mechanical thermal analyser (DMTA 2980, TA Instruments, USA) in dual cantilever bending mode at six different frequencies (from 1 to 200 Hz) and with a constant strain of 0.1% and a temperature range from −120°C to 80°C. The heating rate was 2°C/min. Each sample was cut to a small rectangular bar of dimensions: length 35 mm, width 13 mm and thickness 3 mm. Universal Analysis software was used for data acquisition and analysis. The maximum in tan δ in the plot of tan δ versus temperature was taken as the glass to rubber transition or the glass transition temperature, Tg.

Morphology of Rubber Blends Reinforced with Active Fillers

A JEOL scanning electron microscope was used to perform textural characterization of fractured surfaces of the samples with different content of filler for elastomers based on NBR/EPDM rubber. Each one was fractured cryogenically and submitted to gold sputtering before analysis.

Thermal Ageing of Composites

To investigate the influence of thermal aging on the mechanical properties, the cross-linked samples were placed in an air circulating oven operated at 100°C during 50 h. The retained percentage value of tensile strength and elongation at break were calculated. Tensile properties (modulus at 100 % elongation, tensile strength and, elongation at break) were measured before and after ageing.

Results and Discussion

Compound Curing Characteristics

The curing conversion denotes the extent of the reaction and is determined by measuring the compounds behaviour during the network formation. The most practical method is based on measurement of the isothermal torque vs. time at a given strain. As it is known in accelerated-sulphur-curing systems of EPDM with and without fillers, sulphur cross-linking takes place at the allylic positions independent of the presence of carbon black and oil, Figure 4. For filled compounds, type and content of filler affect the cure characteristics. Lots of functional groups such as hydroxyl, carboxyl, lactone, pyrone, ketone, quinone, and phenol exist on the carbon black surface but the amount is small. The torque-time characteristics for EPDM rubber compounds filled with active and inactive filler are given in our earlier investigation (*34*). Now in the Figure 5 are given the difference between MH and ML as an indirect indication of the crosslink density of the obtained crosslinked materials intended for sealant production. The difference between MH and ML could be used as an indirect indication of the network formation. As can be seen, the torque difference and, thus, the crosslink density increase continuously with increasing carbon black loading, but chalk do not affect crosslinking.

In the Table 1 are summarized the obtained curing characteristics, such as Δ torque (difference between the maximum and minimum torques), scorch time (t_{S2}), and optimum cure time (t_{C90}), for the NBR compounds with different carbon black loading. In the Table 2 are summarized data for NBR/EPDM curing characteristics for different content of carbon black and silica. The values of MH, ML and ΔM increases with the increase of the carbon black loading for all prepared compounds. Both scorch time (t_{S2}) and optimum curing time (t_{c90}) are decreasing with increasing active filler content. In Table 3 are given curing characteristic data for sample based on different precursors. It was estimated that scorch time and optimum curing time decreased with filler loading. On the other hand, minimum and maximum torque increased with filler loading as well. Minimum torque (ML) is directly related to the viscosity of the compounds at the test temperature. The maximum torque (MH) is closely related to the modulus of the obtained crosslinked materials.

In the Figure 6 are given DSC curves for sample with 140 phr of carbon black, and 60 phr of chalk upon heating from -80°C to 170°C. Glass transition temperature was -51.86°C. The crystalline melting point of EPDM macromolecules was registrated at -20.35 °C.

Physico-Mechanical Properties of Composites before and after Aging

Heat aging stability is a hallmark of EPDM rubber and a requirement for use in automotive, roofing and other applications. Retention of physical properties after accelerated heat aging is a typical specification for such long-term uses. In the Table 4 are given some properties of EPDM composites with carbon black (CB) and chalk before ageing (BA) and after ageing (AA) data of tensile properties, and hardness for EPDM composites. As mineral filler, $CaCO_3$ has been used only as bulking agent mainly reducing the cost of rubber products. The main structural parameter of elastomers responsible for their thermal stability is the energy of valence bonds in the macromolecule skeleton. It was estimated that for all samples the hardness was increased with increased filler loadings. The enhanced thermal stability is evidenced from the tensile strength and elongation at break measurements but after aging there is evident the reductions of some properties. An understanding of gas sorption and transport is useful for application to membrane-based gas separation operations, for evaluating EPDM as a potential material for bar-

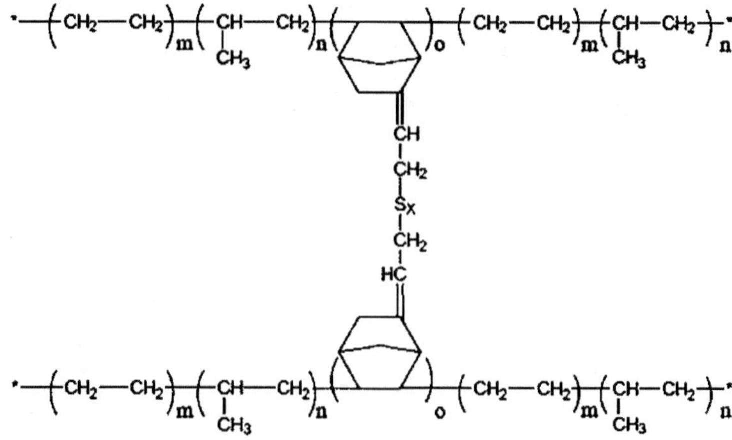

Figure 4. The network structure of obtained samples based on sulfure cured poly(ethylene-co-propylene-co-2-ethylidene-5-norbornene).

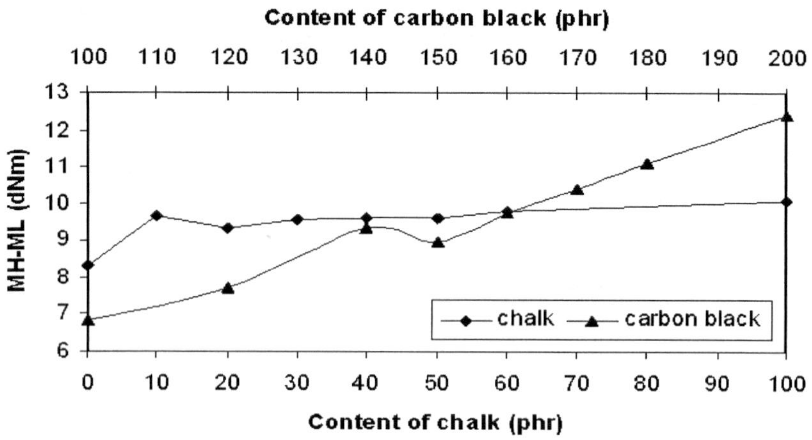

Figure 5. Effects of filler loading on torque difference MH-ML for samples based on EPDM rubber, carbon black nanoparticles and inactive filler chalk.

rier and sealing applications and for predicting the extent of oxidative aging when aging reactions are fast in comparison to oxygen diffusion.

Visco-Elastic Behavior of EPDM Elastomers

Compared to the gum system, the filled system has higher E′ value over the whole range of temperatures throughout all the three regions. Filled system has a lower glass-transition temperature (α) compared to that of the gum system. In the filled system, the shift of the glass rubber transition (Tg) to the lower temperature occurs concomitant with decrease in the tan δ peak height. Oil shifts the peak to low temperatures by penetrating between molecular chain segments and, thus, increasing the free volume. From dynamic mechanical measurement was also

Table 1. Curing characteristics of NBR compounds filled with different content of carbon black

NBR/Carbon black (phr)	*ΔM (Nm)*	t_{S2} *(s)*	t_{C90} *(s)*	*CRI* *(s^{-1})*
100/0	5,87	156	270	0.88
100/20	6,33	120	210	1.11
100/40	8.36	90	180	1.11
100/60	8.87	66	168	0.98
100/80	8.59	54	150	1.04

Table 2. Curing characteristics of NBR/EPDM (80/20 w/w) compounds with different content of carbon black and silica

Carbon black/Silica (phr)	*ΔM MH-ML (Nm)*	t_{S2} *(s)*	t_{C90} *(s)*	*CRI* *(s^{-1})*
70/0	7,13	103	233	0,8
60/10	4.85	150	324	0.57
50/20	4,18	114	264	0.66
40/30	4.29	156	300	0.69
35/35	3,44	138	444	0,32
30/40	3.16	96	228	0.75
20/50	1,90	108	660	0,18
0/70	2,03	12	600	0,17

Table 3. Curing characteristics of NBR/EPDM compounds with different network precursor content filled with 50 phr carbon black

NBR/EPDM	*ΔM (Nm)*	t_{S2} *(s)*	t_{C90} *(s)*	*CRI* *(s^{-1})*
80/20	6.78	60	192	0.75
60/40	5.87	60	162	0.98
50/50	7.01	72	526	0.22
40/60	5.65	60	600	0.18
20/80	3.86	48	96	2.08

observed the second broader tan δ peak (α′) centered on −14°C in the case of filled system. Filled polymer constitutes a system with a complex structure of at least two components: (a) the adsorbed hard rubber (adsorbed within, and between, the filler aggregates and the hard rubber shell surrounding the carbon black); and (b) the bulk rubber. The inclusion in a viscoelastic material not only plays

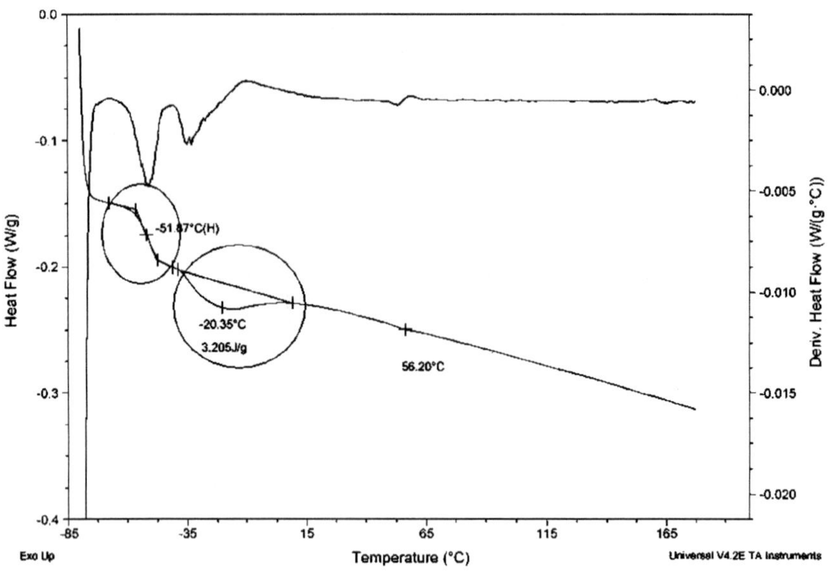

Figure 6. DSC curve for crosslinked EPDM sample filled with 140 phr carbon black and 60 phr chalk

the role of a stress raiser, but also changes the time-dependent behavior of the bulk material. Thus, the adsorbed hard-immobilized rubber causes a perturbed relaxation response and the segmental dynamics of this layer is restricted. By using the time-temperature superposition principle it is possible to measure the viscoelastic data of a polymeric system at different temperatures over a limited period of time/frequency and construct composite curves that extend the effective time scale/the dynamic frequency range to a large extent. Dynamic mechanical behavior of a gum and highly filled EPDM composites was precisely determined over a wide range of frequencies and temperatures. Chemical reaction between the filler and elastomer and occlusion of the rubber macromolecules within the pores of the filler particles effect the mechanical response and final morphology under a given set of conditions of temperature and filler loading level. The addition of fillers to rubber compounds has a strong impact on the static and dynamic behavior of rubber samples. In-rubber structure was interpreted as a combination of the effects caused by the real carbon black structure in the in-rubber state and the filler–polymer interaction. Due to this phenomenon, the effective volume of the filler is increased considerably and should of course enhance the high strain modulus significantly. From dynamic-mechanical measurements it was estimated that with increasing carbon black content elastomer stiffness increases (Figure 7). Carbon black has a very irregular surface, which makes the rubber reinforcement particularly effective. This is a result of the increase in the surface area to volume ratio of the fillers increasing the failure energy.

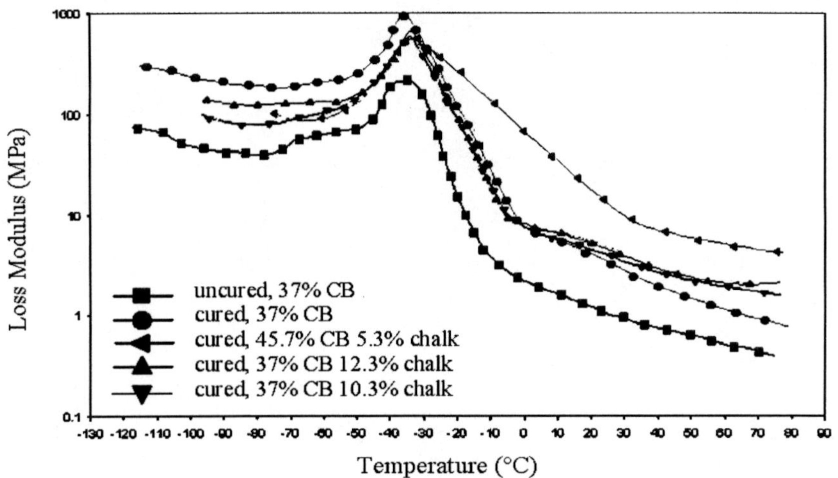

Figure 7. Temperature dependence of bending loss modulus (E") at 1Hz for cured and uncured samples with different content of fillers.

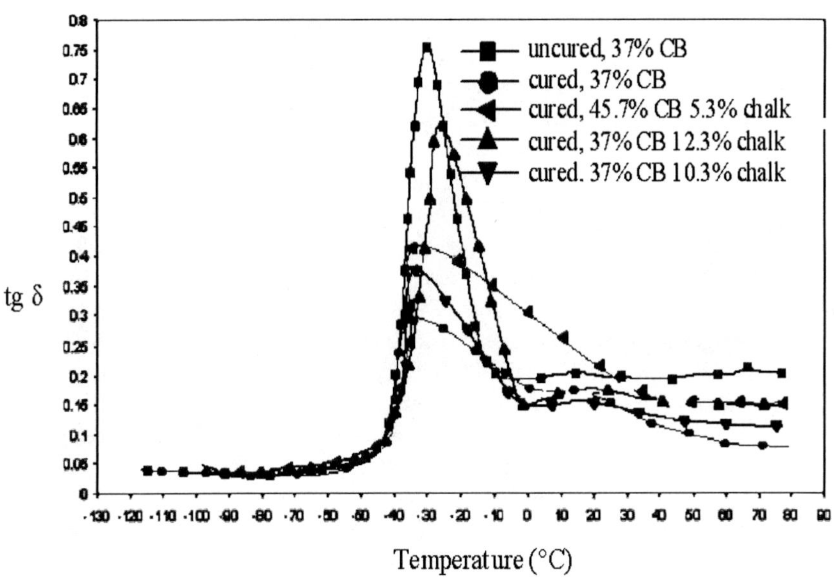

Figure 8. Dependence of bending loss tangent on temperature at frequency 1 Hz for samples with different filler loading

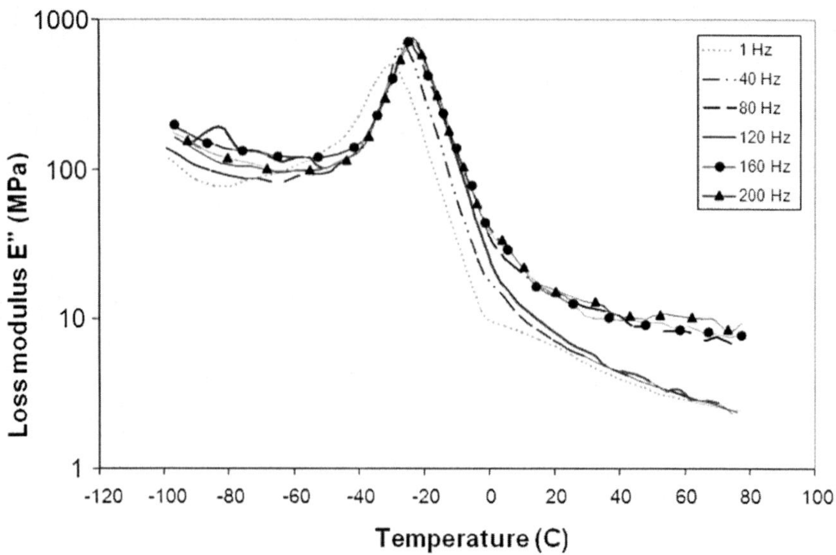

Figure 9. Temperature dependence of bending loss modulus (E") at different frequencies for sample with 37 % CB and 21,8 % of chalk.

Damping properties of the materials which are related to the free volume change are discussed through examination of both height and width of tan δ curves. The tan δ values decrease when the free volume change becomes smaller. Figure 8 displays the damping properties (tan δ) for filled unfilled cured and uncured rubbers. In the Figure 9 can be seen that chalk loading cause better damping properties under high frequencies. Decreasing of peak tan δ values with the increase of particle volume fraction could be explained as smaller mobility of entrapped rubber segments anchored to rigid particle surfaces in the transition region from glassy to rubbery phase. The temperature- and frequency-insensitive damping and elasticity are of significance to extend the availability in the industrial uses. There exists considerable demand for "damping materials" to suppress vibration and noise in industrial fields. The dynamic mechanical analysis of the EPDM rubbers showed that the addition of the filler has no significant effect on the glass transition temperature, *Tg*. However, on increasing filler loading, the intensity of tan δ curve decreases. This has been explained in terms of filler-polymer interaction and inter-aggregate interactions by the concept of formation of interphase layer between the filler and the polymer matrix. The increase of the filler content will persuade the increase of modulus and reduction of the ultimate tensile strength. It is necessary to find a compromise between low friction and tensile strength. The addition of a small percentage of non-reinforcing fillers reduces friction without significant detriment to the modulus and tensile strength. In the reinforced samples, the shift of the glass-rubber transition (*Tg*) to the lower temperature is accompanied with decrease in the tan *δ* peak height. Elastic properties of crosslinked materials can result in the development of tensile stresses when subjected to compression or shear stress. Hence it is necessary to find a careful compromise between low friction and tensile strength. The

Table 4. Physical properties before ageing (BA) and after thermal ageing (AA) for EPDM elastomers filled with different content of CB and 20 phr chalk, and samples with different content of chalk and 140 phr CB

	Hardness (Sh °A)		*Tensile strength (N/mm²)*		*Elongation at break (%)*	
	BA	*AA*	*BA*	*AA*	*BA*	*AA*
100 phr CB	45.7	50	10.34	9.18	1065	918
120 phr CB	53.7	56	10.11	8.96	880.9	760.6
140 phr CB	59.7	59.7	9.05	8.82	799.3	727.6
160 phr CB	65	65.3	8.37	8.31	609.2	549.3
180 phr CB	71	71.3	8.90	8.61	494.9	426.4
200 phr CB	75	75.3	9.57	8.34	315.1	246.2
0 phr chalk	57.7	58.3	9.64	9.95	446.7	397.3
20 phr chalk	59	59.7	9.05	8.82	799.3	727.6
40 phr chalk	61	61.7	8.25	8.17	727.2	686.7
60 phr chalk	62	63	7.36	7.39	726.5	683.6
100 phr chalk	63	63.3	6.43	6.36	669.8	598.7

addition of a small percentage of fillers can reduce friction without the significant detriment to the modulus and the tensile strength. The modulus of a material is related to the hardness at a Figure 10. Materials with a higher modulus reduce the strain for a given stress; hence for a given pressure, acting over a given area, they will strain less and not reach their elongation at break. Harder materials tend to have a higher modulus. Therefore for high pressure applications one may choose to select fillers that will increase the modulus to a greater degree than they will increase hardness.

Properties of Elastomers Based on Rubber Blends

The main structural parameter of elastomers responsible for their thermal stability is the energy of valence bonds in the macromolecule skeleton. The cross-linking of elastomer generally increases its thermal stability, mainly due to the limitation of fluctuation amplitude of macromolecules. The effects of the carbon black and silica filler addition on mechanical properties of the cross-linked EPDM/NBR rubber blends before and after thermal aging are shown in Figures 11121314. It is obvious that the 200% and 300% modulus decreased with increasing silica loading in the rubber blends. The hardness of all crosslinked samples increases with increasing silica contents (Fig. 14) which can be attributed to the increasing cross-linking density after thermal aging. Values of tensile strength, before and after aging decreased with increasing silica content. The tensile strength is a complex function of the nature and type of crosslinks, crosslink densities, the chemical structure of the used elastomers and the changes associated with degradation. It well known that when an elastomer is deformed by an external force, part of the input energy is stored elastically in the chains and is available (released upon crack growth) as a driving force for fracture. The remainder of the energy is dissipated through molecular motions into heat, and in this manner, is made unavailable to break chains. At high crosslink levels, chain motions become restricted, and the dense network is incapable of dissipating much energy. This results in relatively easy, brittle fracture at low elongation.

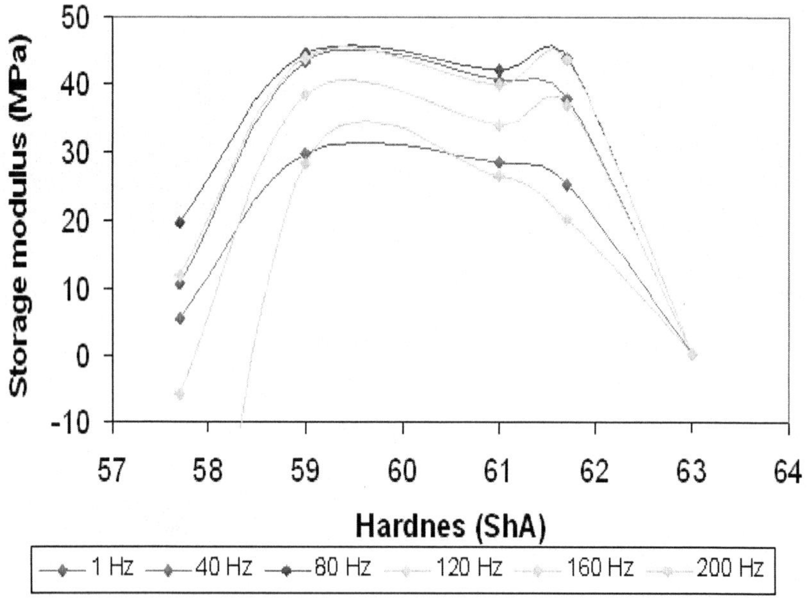

Figure 10. Correlation of storage modulus obtained at different frequencies of dynamic deformation at 25°C with hardness for EPDM composites

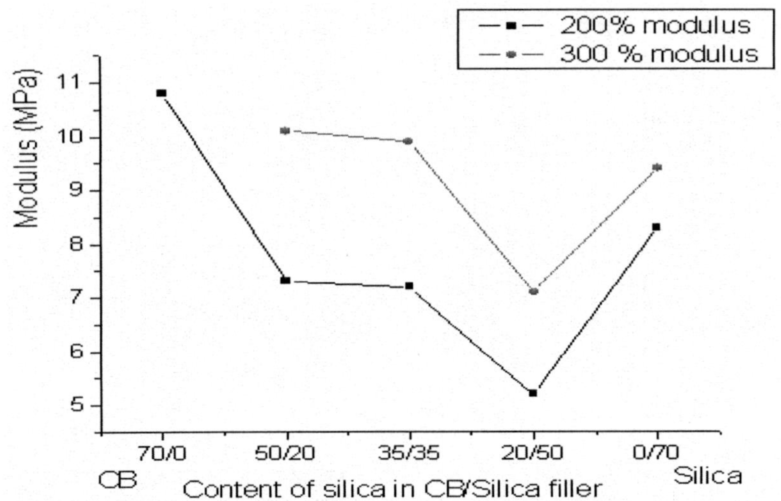

Figure 11. The modulus of elastomers based on rubber blend with different content of active fillers silica and carbon black.

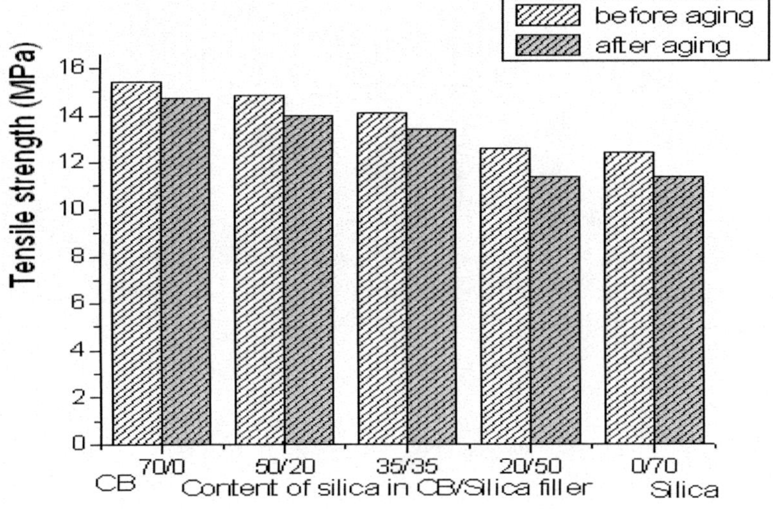

Figure 12. The tensile strength of EPDM/NBR elastomers with different content of active fillers silica and carbon black.

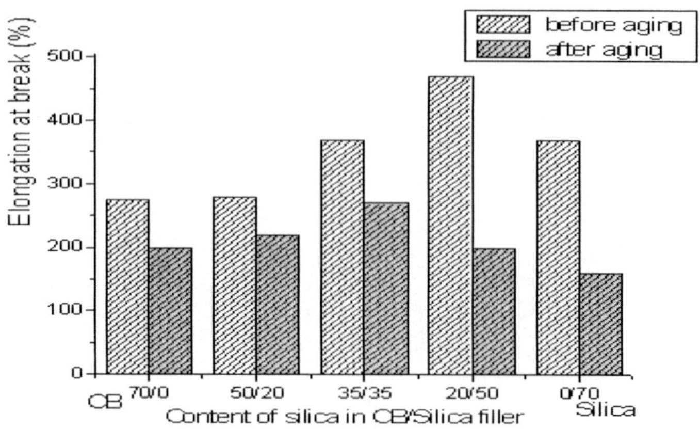

Figure 13. The elongation at break of NBR/EPDM rubber blends with different content of silica and carbon black.

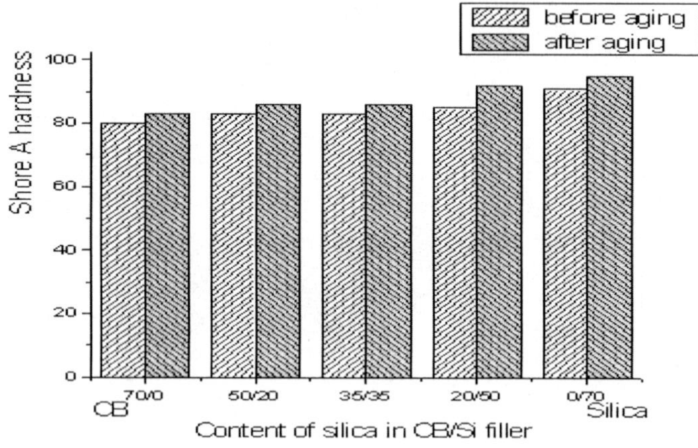

Figure 14. The hardness of elastomers with different content of silica and CB

Morphology of Elastomers Based on Different Active Filler

The network precursors was chosen partly because of their differences in relative polarity and solubility parameters, as well as the ease of distinguishing them under electron microscopy through their electron density contrast. In the Figures 1516171819 are given SEM images for unfilled and CB filled elastomers based on one network precursor. In the Figures 20 and 21 are given SEM images of elastomers based on two network precursor filled with CB and/or silica.

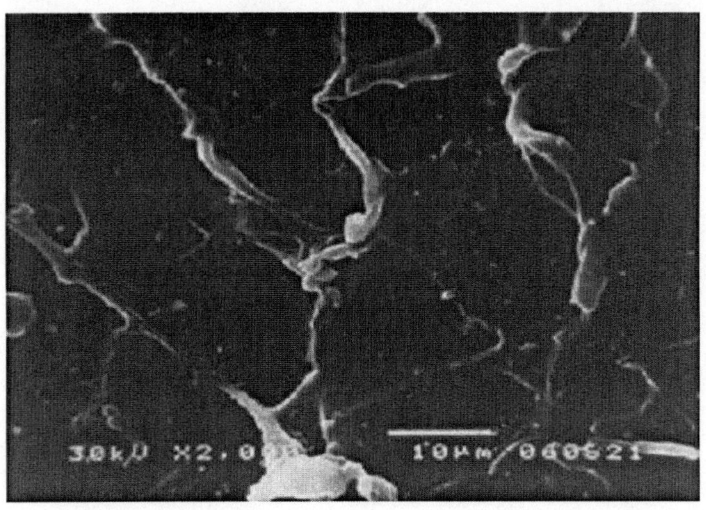

Figure 15. SEM micrograph of elastomers based on unfilled NBR rubber (magnification X=2000).

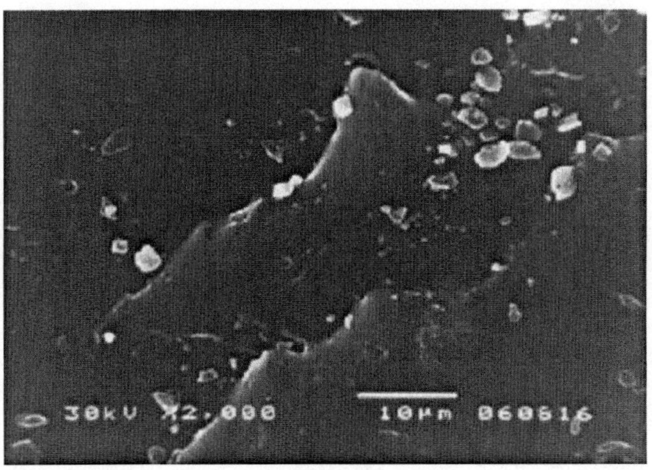

Figure 16. SEM micrograph of elastomers based on NBR rubber filled with 50 phr carbon black (magnification X=2000).

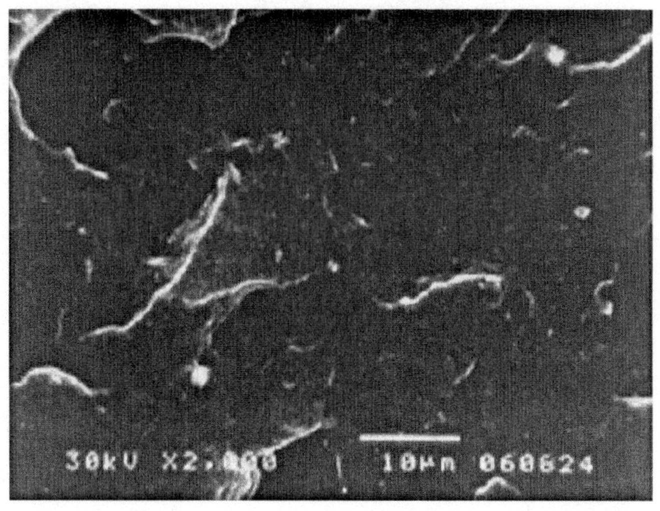

Figure 17. SEM micrograph of elastomers based on unfilled EPDM rubber (magnification X=2000).

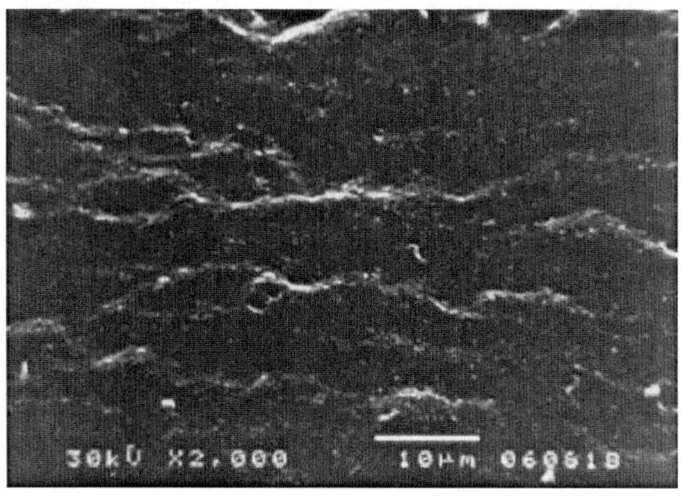

Figure 18. SEM micrograph of elastomers based on EPDM rubber filled with 50 phr carbon black (magnification X=2000).

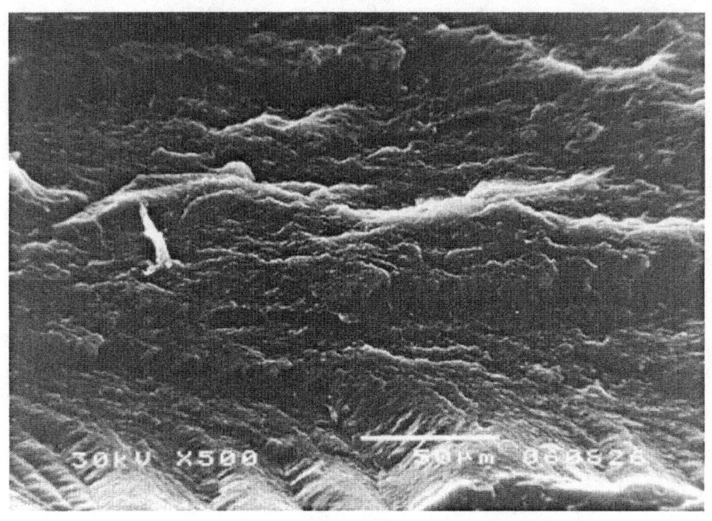

Figure 19. SEM micrograph of NBR/EPDM rubber blend filled with 70 phr carbon black (magnification X=500).

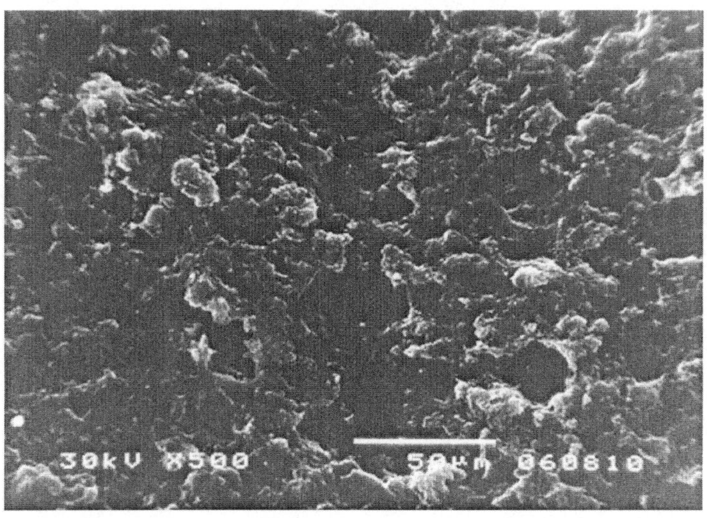

Figure 20. SEM micrograph of elastomer based on NBR/EPDM rubber blend filled with 50 phr carbon black and 20 phr of silica (magnification X=500).

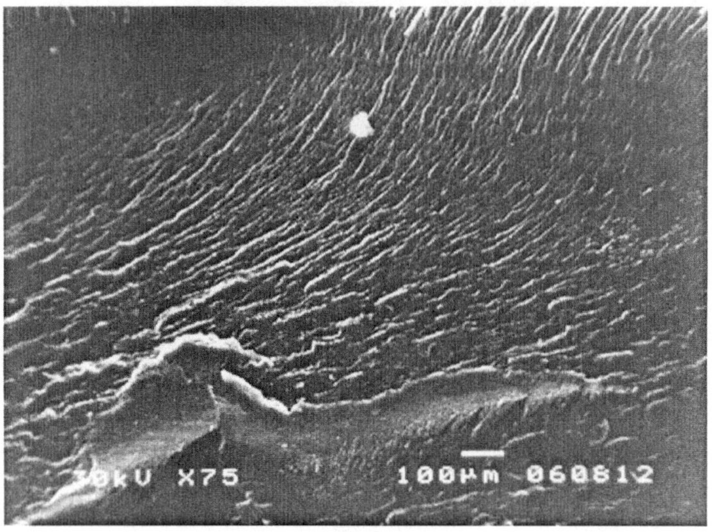

Figure 21. SEM micrograph of elastomer based on NBR/EPDM rubber blend filled with 70 phr of silica (magnification X=75).

Conclusions

The present applicative study was undertaken to give a more detailed analysis of the influence of fillers (carbon black and chalk) to the properties of elastomer based on EPDM rubber as network precursor. In the first step of investigation it was estimated that the ratio of active and inactive filler loading was the prime factor for determining the dynamic-mechanical behaviour of obtained materials. It was estimated for that the samples with high filler loading have higher E′ value over the whole range of temperatures for ealstomers based on EPDM. It was found that the addition of chalk in small concentrations, has a positive contribution to modulus, and does not change tensile strength and elongation at break. The tan δ position shifts towards lower temperature. The higher temperature maximum in tan δ is due to the adsorbed hard rubber around the carbon black aggregates. Segmental mobility of the interphase layer of the polymer adsorbed on the filler surface is restricted and shifted towards the higher temperature. In the second step of investigation the samples with enhanced ozone and weather resistance were prepared by blending two network precursors (NBR and EPDM) and filled either with high abrasion furnace carbon black (46 nm) or with precipitated silica (22 nm) or its combination. It is observed that the addition of carbon black to rubber blend tends to build torque more rapidly than silica. It was found that the optimum cure time significantly decreased with NBR content. It was estimated that unfilled elastomers based on NBR rubber have very low tensile strength because of no self-reinforcing effect, but when used in combination with carbon black nanoparticles, materials with excellent mechanical properties were obtained. The enhanced thermal stability of filled samples is evidenced from the tensile strength and elongation at break measurements but after aging there is noticed the reductions of some mechanical properties. Carbon black also improves the ageing resistance of the rubber material. This is due to the ability of carbon black to absorb UV light, which otherwise would be absorbed by the polymer and initiate photo-oxidative reactions leading to deterioration. It was estimated considerable performance enhancement of rubber blends by reinforcing with dual active fillers.

References

1. Tariq, Y.; Shamshad, A.; Fumio, Y.; Keizo, M. Radiation vulcanization of acrylonitrile butadiene rubber with polyfunctional monomers. *React. Funct. Polym.* **2002**, *53*, 173–181.
2. EL-Nashar, D. E.; Gommaa, S.; E-Messieh, A. Study of electrical, mechanical, and nanoscale free-volume properties of NBR and EPDM rubber reinforced by bentonite or kaolin. *J. Polym. Sci., Part B: Polym. Phys.* **2009**, *47*, 1825–1838.
3. Zaharescua, T.; Meltzer, V.; Pincu, E.; Jipa, S. Thermal study on binary blends of ethylene–propylene elastomers and acrylonitrile–butadiene rubber. *Polym. Bull.* **2007**, *58*, 683–689.
4. Mathur, G.; Pandey, K.; Setu, D. *Polym. Eng. Sci.* **2005**, *45* (9), 1265–1276.
5. Abou, M.; Zeid, S. Radiation effect on properties of carbon black filled NBR/EPDM rubber blends. *Eur. Polym. J.* **2007**, *43* (10), 4415–4422.
6. Rajeev, R.; De, S. Crosslinking of rubbers by filler. *Rubber Chem. Technol.* **2002**, *75* (3), 475–509.
7. Frohlich, J.; Niedermeier, W.; Luginsland, H. D. The effect of filler–filler and filler–elastomer interaction on rubber reinforcement. *Composites, Part A* **2005**, *36*, 449–460.
8. Allegra, G.; Raos, G.; Vacatello, M. Theories and simulations of polymer-based nanocomposites: From chain statistics to reinforcement. *Prog. Polym. Sci.* **2008**, *30* (7), 683–731.
9. Coran, Y. Chemistry of the vulcan ization and protection of elastomers: A review of the achievements. *J. Appl. Polym. Sci.* **2003**, *87* (1), 24–30.
10. Markovic, G.; Radovanovic, B.; Cincovic-Marinovic, M.; Budinski-Simendic, J. Investigations of SBR/CSM blends reinforced by carbon black. *Kautsch. Gummi Kunstst.* **2006**, *59* (5), 251–255.
11. Lazic, N.; Budinski-Simendic, J.; Petrovic, Z.; Plavsic, M. Modification of dynamic properties of SBR rubber composites with silica fillers. *Mater. Sci. Forum* **2006**, *518*, 417–422.
12. Cassagnau, P. Payne effect and shear elasticity of silica filled polzmers in concentrated solutions and in molten state. *Polymer* **2003**, *44*, 2455–2462.
13. Medalia, A. Morphology of aggregates: VI. Effective volume of aggregates of carbon black from electron microscopy; Application to vehicle absorption and to die swell of filled rubber. *J. Colloid Interface Sci.* **1970**, *32*, 115–131.
14. Wolff, S. Chemical aspects of rubber reinforcement by fillers. *Rubber Chem. Technol.* **1996**, *69*, 325–346.

15. Markovic, G.; Marinovic-Cincovic, M.; Valentova, H.; Ilavsky, M.; Radovanovic, B.; Budinski-Simendic, J. Curing characteristics and dynamic mechanical behaviour of reinforced acrylonitrile-butadiene/chlorosulfonated polyethylene rubber blends. *Mater. Sci. Forum* **2005**, *494*, 475–481.
16. Agullo, N.; Borros, S. Qualitative and quantitative determination of the polymer content in rubber formulations. *J. Therm. Anal. Calorim.* **2002**, *67*, 513–522.
17. Markovic, G.; Devic, S.; Marinovic-Cincovic, M.; Budinski-Simendic, J. Influence of carbon black on reinforcement and gamma-radiation resistance of EPDM/CSM CR/CSM rubber blends. *Kautsch. Gummi Kunstst.* **2009**, *62* (6), 299–305.
18. Essawy, H.; El-Nashar, D. The use of montmorillonite as a reinforcing and compatibilizing filler for NBR/SBR rubber blend. *Polym. Test.* **2004**, *23*, 803–807.
19. Hanafi, I.; Tan, S.; Poh, B. T. Curing and mechanical properties of nitrile and natural rubber blends. *J. Elastomomers Plast.* **2001**, *33* (4), 251–262.
20. Botros, S. H.; Tawfic, M. L. Compatibility and thermal stability of EPDM-NBR elastomer blends. *J. Elastomomers Plast.* **2005**, *37* (4), 299–317.
21. Leblanc, J. L. Rubber–filler interactions and rheological properties in filled compounds. *Prog. Polym. Sci.* **2002**, *27* (4), 627–687.
22. Vilgis, T. A. Time scales in the reinforcement of elastomers. *Polymer* **2005**, *46*, 4223–4229.
23. Wolf, S. Chemical aspects of rubber reinforcement by fillers. *Rubber Chem. Technol.* **1996**, *69*, 325–346.
24. Zaborski, M.; Donnet, J. B. Activity of fillers in elastomer network of differente structure. *Macromol. Symp.* **2003**, *194*, 87–99.
25. Rajeev, R. S.; De, S. K. Crosslinking of rubbers by fillers. *Rubber Chem. Technol.* **2002**, *75*, 475–509.
26. Gerspacher, M. Advanced CB characterizations to better understand polymer-filler interaction. *Kautsch. Gummi Kunstst.* **2009**, *5*, 233–239.
27. Donnet, J. B. *Kautsch. Gummi Kunstst.* **1994**, *47*, 628–632.
28. Markovic, G.; Radovanovic, B.; Marinovic-Cincovic, M.; Budinski-Simendic, J. The effect of accelerators on curing characteristics and properties of natural rubber/chlorosulphonated polyethylene rubber blend. *Mater. Manuf. Processes* **2009**, *24*, 1224–1228.
29. Eid, M. A.; El-Nashar, D. E. Filling effect of silica on electrical and mechanical properties of EPDM/NBR blends. *Polym.−Plast. Technol. Eng.* **2006**, *45* (6), 675–684.
30. Essawy, H.; El-Nashar, D. The use of montmorillonite as a reinforcing and compatibilizing filler for NBR/SBR rubber blend. *Polym. Test.* **2004**, *23* (7), 803–807.
31. Zaharescu, T.; Meltzer, V.; Vîlcu, R. Thermal properties of EPDM/NR blends. *Polym. Degrad. Stab.* **2000**, *70* (3), 341–345.
32. Ghosh, A. K.; Basu, D. K. Co-vulcanization of acrylonitrile−butadiene rubber and ethylene−propylene−diene rubber blends. *Kautsch. Gummi Kunstst.* **2003**, *56* (3), 101–109.

33. El-Sabbagh, S. H. Compatibility study of natural rubber and ethylene–propylene diene rubber blends. *Polym. Test.* **2003**, *22* (1), 93–100.
34. Milić, J.; Aroguz, A.; Budinski-Simendić, J.; Radičević, R.; Prendzov, S Morphology and viscoelastic properties of sealing materials based on EPDM rubber. *J. Microsc. (Oxford, U.K.)* **2008**, *232* (3), 580–584.

Simulation and Modeling in Polymer Science

Chapter 13

Molecular-Level Deformations in Auxetic Organic Networked Polymers

Joseph N. Grima,* Ruben Gatt, Daphne Attard, and Richard N. Cassar

Department of Chemistry, University of Malta, Msida MSD 2080, Malta
***e-mail: joseph.grima@um.edu.mt**

Auxetic materials exhibit the unusual property of getting fatter when stretched and thinner when compressed, i.e. exhibit a negative Poisson's ratio. Here we simulate and analyze three types of 2D networked polymers two of which are polyphenylacetylene polymers and the third is constructed from calix[4]arene building blocks. We simulate the molecular level deformations that take place when these systems are stretched in various directions in an attempt to obtain an insight into the molecular level deformations which result in auxeticity. The results of these simulations are then used to explain why these polymers are auxetic for loading in certain directions but not in others.

Introduction

Auxetic materials exhibit the unusual property of getting fatter when stretched (rather than thinner) and getting thinner when compressed (rather than fatter), thus having a negative Poisson's ratio (*1*, *2*). These materials have been shown to exhibit various enhanced properties when compared to conventional materials, ranging from an enhanced indentation resistance to the ability of forming dome-shaped surfaces and are thus superior to conventional materials in various practical applications. In recent years there have been various reports in literature describing advances in the field of auxetics relating to the design, modeling, synthesis and/or discovery of auxetic materials (*1–48*). One area which has seen significant advances *vis-à-vis* auxetic behavior is that of auxetic polymers (*1–25*) with Griffin *et al.*'s synthesis work on liquid crystalline polymers (*3*, *4*). Examples of auxetic polymers include Baughman *et al.*'s three-dimensional 10,3-*b* auxetic polymeric networks (*5*), polymeric foams and microporous polymers (*16–25*) and various classes of organic 2D networked polymers which have been proposed to be potentially auxetic (*1, 3–13*). This paper discusses three of these 2D networked polymers namely:

1. the polyphenylacetylene networks commonly referred to as 'reflexynes' originally proposed by Evans *et al.* (*1, 6, 7, 12, 13*) (see Fig. 1a);
2. the polyphenylacetylene networks 'polytriangles' networks (see Fig. 1b) proposed by Grima and Evans respectively (*8, 9, 13*); and
3. the 'polycalixes' built using calix[4]arene building blocks (see Fig. 1c) also proposed by Grima *et al.* (*8, 10, 11, 13*).

These three classes of organic networked polymers have all been designed to resemble the shape of known auxetic macrostructures (compare Fig. 1 with Fig. 2). For example, the reflexynes were designed with the aim of achieving molecular-level auxetics which resemble the well known auxetic re-entrant honeycomb structures (see Fig. 2a) where the ribs of the honeycombs are represented by acetylene chains whilst the joints are constructed from phenyl rings. In such systems, the 're-entrant connectivity' is achieved by having 1,2,3-subsituted phenyl rings (to allow angles of 60° between adjacent acetylene chains). Similarly, the polytriangles were designed to mimic a structure made from connected triangles which rotate relative to each other (see Fig. 2b) whilst the polycalixes where meant to imitate the behavior of a folded 'egg-rack' macrostructure (see Fig. 2c).

Although there has been various studies which suggest that these three systems are predicted to exhibit some degree of auxetic character (*1, 3–12*), it was only recently that an attempt was made to compare them together (*13*), a study which suggested some very interesting results. For example, it was found that despite the fact that the reflexynes are probably the systems which have been so closely associated with negative Poisson's ratios in view of their characteristic re-entrant geometry, they are actually the least auxetic of the three classes of honeycombs considered in the study. In fact, on analyzing the in plane mechanical properties of the reflexynes (i.e. the plane of the re-entrant honeycombs), it was found that these systems only exhibited auxetic behavior for loading approximately on-axis with the Poisson's ratio being positive for loading in most of the other directions. Instead, it was found that the 'polytriangles' can be regarded as showing the highest degree of auxeticity since it was found that 'polytriangles' polymeric systems are always auxetic for loading in any direction in the plane of the triangles with the Poisson's ratio tending to -1 for the systems with larger triangles. These systems were also found to be isotropic in the plane of the triangles meaning that the same extent of auxeticity was measured irrespective of the direction of loading.

These recent findings highlight the fact that these 2D networked polymers need to be studied in greater detail. For example, there is still little information regarding the molecular level deformations as a result of uniaxial stresses applied both on- and off-axis, an issue which will be addressed in the present study. In particular, in this work we use force-field based simulations to predict the atomic level deformations as a result of stresses applied on-axis or off-axis in an attempt to explain the values of the Poisson's ratios obtained.

Methodology

Three different classes of polymers were examined, namely reflexynes (**1**), polytriangles (**2**) and polycalixes (**3**) and for each of these classes, three different systems (homologues) were studied (**A-C**). These nine polymer systems, illustrated in Fig. 1, were studied using force-field based simulations using the commercially available software package Cerius^2, V4.2 (Accelrys Inc., San Diego, USA) with the aim of simulating the in-plane on-axis and 45° off-axis Young's moduli and Poisson's ratios and the way these systems deform when they are uniaxially mechanically loaded (stretched or compressed) on-axis or at 45° off-axis. These simulations were performed in an attempt to obtain a better insight into the deformation mechanisms of these systems.

All systems were represented as crystals (with a P1 symmetry) so as to represent infinite systems. The crystals were oriented in the global *XYZ* space in such a way that the [001] direction was always parallel to the Z-axis with the [010] direction lying in the YZ-plane. When systems are aligned in this way within the *Cerius^2* environment, then one ensures that the plane of the networks remains parallel to the YZ-plane. Furthermore, this alignment enables a direct comparison with earlier work by Grima *et al.* (*13*) performed using Materials Studio (Accelrys Inc., San Diego, USA). For the reflexyne and polytriangle systems, the networks were allowed to stack freely 'on each other' in the third direction, while for the polycalixes they were allowed to stack freely 'inside one another' in the third direction. The energy expression was set up using the PCFF (*49*) force-field with its standard parameters and settings, a force-field which is adequately parameterized to model our systems (*13*). Non-bond terms (Van der Waals and Coulombic interactions) were added using the EWALD summation technique (*50*). This technique was used since it gives lower cut-off errors for modeling of crystals (*51*) when compared to simpler cut-off techniques such as the direct or spline cut-off. Energy minimization was performed using SMART compound minimizer as implemented in the OFF method Simulation Engine within the *Cerius^2* with default high convergence criteria settings which include an energy difference of 1×10^{-4} kcal/mol, a RMS displacement of 1×10^{-5} Å and a maximum displacement of 5×10^{-5} Å. During the minimization, all cell parameters were set as variables, *i.e.* no constraints on the shape and size of the unit cell were applied. The 6 x 6 stiffness matrix **C** of the minimum energy conformation of the systems was calculated from the second derivative of the potential energy function since

$$c_{ij} = \frac{1}{V} \frac{\partial^2 E}{\partial \varepsilon_i \partial \varepsilon_j} \quad i, j = 1, 2, \ldots, 6$$

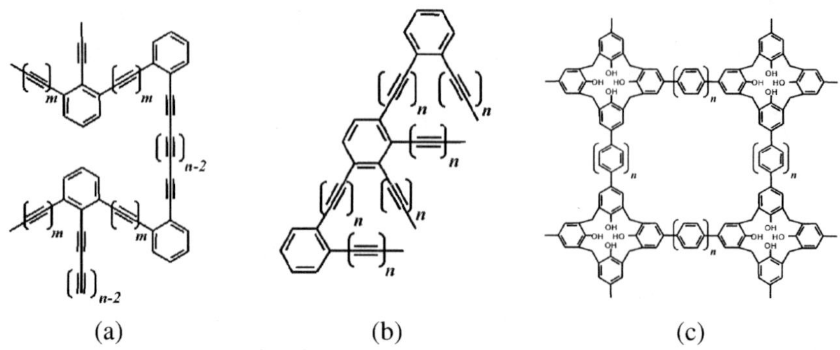

*Figure 1. An idealized 2D representation of the systems modeled: (a) the reflexynes **1A** – **C** where for **1A**, **1B** and **1C** (m,n) = (1,4), (1,5) and (1,6) respectively, (b) the polytriangles **2A** – **C** where for **2A**, **2B** and **2C**, n = 3, 4 and 5 respectively, and (c) the polycalixes **3A** – **C** where for **3A**, **3B** and **3C**, n = 0, 1 and 2 respectively.*

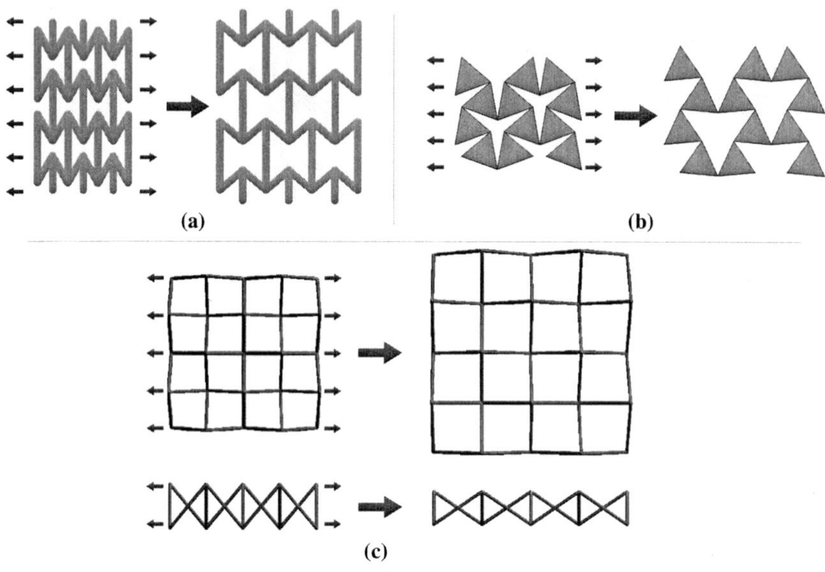

Figure 2. (a-c) An illustrations of the idealized models/mechanisms in operation: (a) auxeticity from hinging re-entrant honeycombs; (b) auxeticity from rotating hinging triangles; (c) auxeticity from the 'egg rack' / 'opening of an umbrella' mechanism.

where c_{ij} is an element of the stiffness matrix **C**, E is the energy expression, V is the volume of the unit cell and ε_i and ε_j are strain components where $\varepsilon_1 = \varepsilon_x$, $\varepsilon_2 = \varepsilon_y$, $\varepsilon_3 = \varepsilon_{yz}$, $\varepsilon_6 = \varepsilon_{xy}$. The in-plane on-axis and off-axis Poisson's ratios and moduli of these systems were calculated from the compliance matrix $\mathbf{S} = \mathbf{C}^{-1}$. In particular we note that for the YZ plane i.e. the plane of the networks, the on-axis moduli and Poisson's ratio are given by:

$$\text{Young's moduli: } E_y = \frac{1}{s_{22}}, \; E_z = \frac{1}{s_{33}}$$

$$\text{Poisson's ratios: } \nu_{yz} = -s_{32}E_y = -\frac{s_{32}}{s_{22}}, \; \nu_{zy} = -s_{23}E_z = -\frac{s_{23}}{s_{33}}$$

$$\text{Shear moduli: } G_{yz} = \frac{1}{s_{44}}$$

The off-axis mechanical properties were then obtained using standard axis transformation techniques, as described in (*52*).

In an attempt to obtain a better understating of the molecular level deformations of these polymers when they are subjected to uniaxial stresses we performed additional simulations where the polymers were minimized whilst subjected to uniaxial stresses applied along the *Y* axis, *Z* axis and at 45° off-axis. All stresses were applied in tension, usually within the range of 0 to 10% of the Young's moduli calculated at zero strain in that particular direction of stretching. In Cerius[2], such stresses may be applied by entering the components of stress tensor $\boldsymbol{\sigma}$ where for stresses of magnitude σ applied in the *Y*, *Z* or 45° off-axis in the *YZ* plane, the following respective stress tensors were used:

Y direction

$$\boldsymbol{\sigma} = \begin{pmatrix} 0 & 0 & 0 \\ 0 & \sigma & 0 \\ 0 & 0 & 0 \end{pmatrix}$$

Z direction:

$$\boldsymbol{\sigma} = \begin{pmatrix} 0 & 0 & 0 \\ 0 & 0 & 0 \\ 0 & 0 & \sigma \end{pmatrix}$$

45° off-axis in YZ plane:

$$\boldsymbol{\sigma} = \frac{1}{2}\begin{pmatrix} 0 & 0 & 0 \\ 0 & \sigma & \sigma \\ 0 & \sigma & \sigma \end{pmatrix}$$

The deformations so obtained were then used to explain the particular values of the Poisson's ratio exhibited by the various polymers.

Results and Discussion

The in-plane on-axis Poisson's ratios, Young's and shear moduli are tabulated in table 1 whilst the in-plane off-axis Young's moduli and Poisson's ratios are reported in figure 3. These results suggest that the results of our simulations compare well to those found in the literature. They also confirm that all systems exhibit some degree of auxeticity, a property which is most predominant in the polytriangles and least in the reflexynes, this being in accordance with the results obtained in a similar study (*13*). Also, it appears, not unexpectedly, that the more "open" networks have significantly reduced densities. This may have some undesirable limitations regarding the practical applications for these networks. This is due to the fact that there might be a problem of empty space at the sub-nano scale that could produce in effect a very high level of surfaces without van der Waals contacts. To rectify this problem, one could include some inert species in the empty spaces which could stabilize these systems. However, such inclusions are likely to have some effect on the mechanical properties of these systems, as was observed in the case of the auxetic zeolite natrolite (*39*, *47*) and thus further work in this regard is required.

Table 1. ***The on-axis mechanical properties of reflexynes (1A-1C), polytriangles (2A-2C) and 'double calixes' (3A-3C) obtained in this study using the PCFF force-field found in Cerius²-OFF compared to values obtained from literature***

	Method	ν_{yz}	ν_{zy}	E_y *(GPa)*	E_z *(GPa)*	G_{yz} *(GPa)*	*Density (g/cm³)*
1A	PCFF – *Cerius²*	-0.37	-0.33	125.90	112.00	5.11	1.21
	Ref. (13)	*-0.41*	*-0.34*	*120.05*	*99.38*	*4.57*	*1.20*
	Ref. (7)	*-0.29*	*-0.29*	*124*	*110*	-	-
	Ref. (6)	-	-	-	-	-	-
1B	PCFF – *Cerius²*	-0.37	-0.41	92.63	101.45	3.63	1.02
	Ref. (13)	-0.37	-0.41	92.68	101.71	3.64	1.02
	Ref. (7)	-0.29	-0.39	95	116	-	-
	Ref. (6)	-0.33	-0.39	94	110	-	-
1C	PCFF – *Cerius²*	-0.31	-0.46	78.40	114.28	2.75	0.93
	Ref. (13)	-0.31	-0.46	78.20	114.25	2.65	0.92
	Ref. (7)	-0.22	-0.42	84	140	-	-
	Ref. (6)	-0.28	-0.44	80	124	-	-
2A	PCFF – *Cerius²*	-0.51	-0.51	52.49	52.51	53.57	0.90
	Ref. (13)	-0.51	-0.51	52.56	52.60	53.52	0.90
	Ref. (9)	-0.83	-0.83	-	-	-	-
2B	PCFF – *Cerius²*	-0.65	-0.65	31.60	31.60	45.64	0.75
	Ref. (13)	-0.65	-0.65	31.51	31.48	45.50	0.75
	Ref. (9)	-0.90	-0.90	-	-	-	-
2C	PCFF – *Cerius²*	-0.75	-0.75	20.06	20.02	39.86	0.64
	Ref. (13)	-0.75	-0.75	20.18	20.12	39.75	0.64
	Ref. (9)	-0.94	-0.93	-	-	-	-
3A	PCFF – *Cerius²*	-0.51	-0.51	18.09	18.09	4.55	1.42
	Ref. (13)	-0.47	-0.47	18.79	18.79	4.61	1.42
	Ref. (9)	-0.51	-0.51	18.10	18.10	-	-
3B	PCFF – *Cerius²*	-0.87	-0.87	3.78	3.79	2.11	1.10

Continued on next page.

Table 1. (Continued). ***The on-axis mechanical properties of reflexynes (1A-1C), polytriangles (2A-2C) and 'double calixes' (3A-3C) obtained in this study using the PCFF force-field found in Cerius²-OFF compared to values obtained from literature***

	Method	ν_{yz}	ν_{zy}	E_y *(GPa)*	E_z *(GPa)*	G_{yz} *(GPa)*	*Density* *(g/cm³)*
	Ref. (*13*)	-0.86	-0.86	3.65	3.65	1.95	1.10
	Ref. (*9*)	-0.87	-0.85	3.78	3.78	-	-
3C	PCFF – *Cerius²*	-0.91	-0.92	1.81	1.82	1.10	0.89
	Ref. (*13*)	-0.95	-0.87	1.67	1.52	0.99	0.89
	Ref. (*9*)	-0.93	-0.88	1.67	1.58	-	-

Let us now attempt to explain these profiles of the Poisson's ratios for the various polymers in terms of the molecular level deformations, a process which will make it possible for us to understand clearly the factors which aid or hinder the presence of a negative Poisson's ratios.

As illustrated in figure 4, in the case of the simulations on the reflexyne polymers which had originally been designed to mimic the re-entrant flexing/hinging honeycombs (see Fig. 5), the simulations suggest that all the three reflexyne structures studied behave, to a first approximation, like 're-entrant flexing/hinging honeycombs' a process which results in auxetic behavior and an increased pore size (see Fig. 4).

In fact, from several measurements of bond lengths and angles on the intitial and deformed structures of (1,4)-reflexyne, whose results are summarised in Table 2, it is evident that for tensile loading (up to 2GPa) in both the *Y* and *Z* directions, the benzene rings remain almost unaffected throughout the deformation process (except for the bond angle at the benzene – vertical acetylene chain junction during loading in the Z-direction). Stretching of the bonds in the inclined acetylene chains (and the vertical chain for stretching in the Z-directions) offers a significant contribution to the overall deformation of the network. In this respect, it is inetersting to note that stretching occurs mostly in the bonds joining the acetylene chain to the benzene ring, a fact which has already been outline elsewhere (*6*). In fact, the fractional change in length of these bonds is around twice as much as that for the other bonds. On the other hand, changes in the bond angles in each acetylene chain are negligible when compared to changes in the angle at the benzene ring – inclined acetylene chain junctions. This suggests that flexing of the acetylene chains is insignificant at the stresses considered and that deformations are mostly focused on hinging at the junctions. This is in accordance to earlier work by Masters and Evans (*48*) which suggests that at the nanolevel, hinging and stretching may play a very important role in the overall deformation of the system, although flexing, which also leads to auxetic behaviour, becomes significant as the length of the shorter chains increases. Also, the fact that these hypothetical networks are able to exhibit a negative Poisson's ratio confirms that it is indeed possible to obtain auxetic behaviour at the nanolevel by downscaling from macro-level structures.

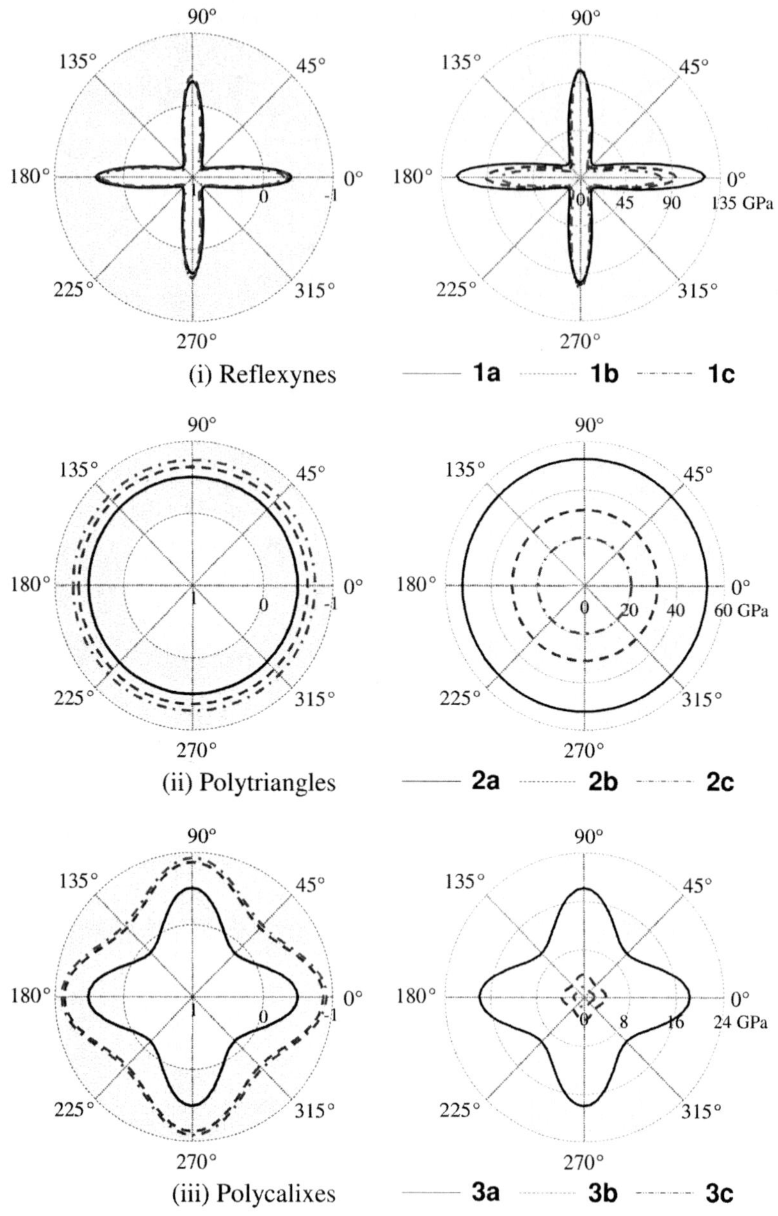

Figure 3. Off-axis plots for the in-plane Poisson's ratios, ν_{yz} and Young's moduli E_y, for each of the network systems considered as obtained by the PCFF force-field.

If we now look at the molecular level deformations that occur when the polymer is stretched at 45° off-axis (Fig. 4c), we note that the deformation profile is completely different to what occurs when loading on-axis thus explaining why the off-axis Poisson's ratio are not negative (a property which has not been given much attention in recent years). In fact, the simulations suggest that the longer vertical arms of all three structures flex significantly more when compared to the side arms which in fact remain almost undeformed. All this suggests that although deformations still occur as a result of flexing/hinging of the acetylene chains, the manner of deformation is not one which is conducive to auxetic behavior since the net result is that the well known 're-entrant mechanism' illustrated in figure 2a does not operate for stretching off-axis thus leading to the complete loss of auxetic character when loading off-axis. This conventional behaviour for loading at certain off-axis angles has already been discussed elsewhwere (*48*) where it has been shown that re-entrant honeycombs exhibit conventional behaviour if they deform primarily through hinging/flexure of the ribs. It should be emphasized that from a structural point of view, such flexing of the long vertical chains (ribs) is easier to induce than flexure of the shorter side chains (ribs) thus explaining the lower Young's moduli when stretching off-axis when compared to stretching on-axis (see Fig. 3). Also, it is interesting to note that the way the network deforms when loaded at 45° off-axis is very similar to the way they behave when sheared on-axis, a property which is characteristic of all hexagonal honeycombs of this form which are known to be weak in shear, thereby limiting, to some extent, the suitability of these systems in practical applications which require high strength.

In the case of the polytriangle structures, as illustrated in figure 6 for the **2B** system which typifies the properties of all three polytriangles studied, the simulations suggest that in whatever direction the stress is applied in-plane, the acetylene chains are observed to flex in such a way that there is a net effect where the triangles in the structure effectively rotate relative to each other. This process results in the opening up of the larger pores of the structure, which pores open in a symmetric manner, a property which may be useful in applications of molecular filtration. This direction-independence of the properties, together with the high negative Poisson's ratios observed, is very significant and makes these systems highly auxetic. Also significant is the fact that for the systems considered here, changes in the size of the triangle affect the Young's modulus to a much higher extent than the Poisson's ratio, i.e. the 'rotating triangles' deformation mechanisms remains the most prominent in all cases. This allows for the possibility to design auxetic polymers having a variety of stiffnesses but without significantly compromising its auxetic behaviour.

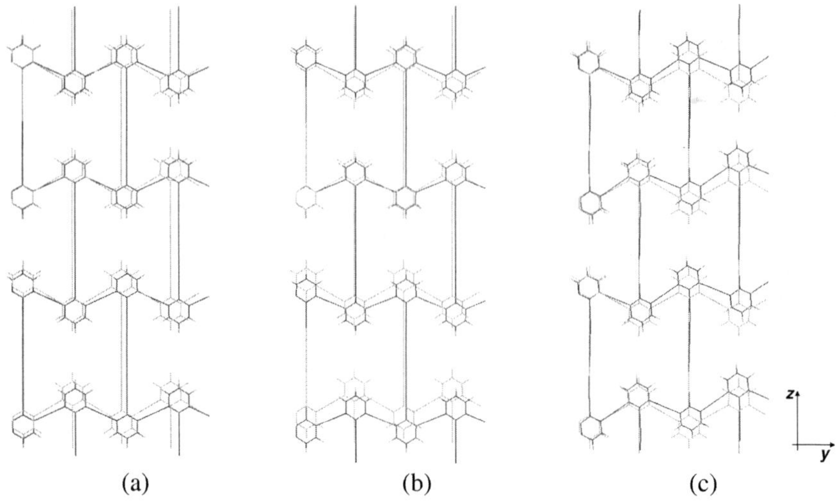

*Figure 4. Deformation mechanism of the 1,5-reflexyne (**1B**) due to a stress (of 6% of the Young's modulus in the direction of the stress) along (a) the Y-, (b) Z- and (c) 45° Z-axis. The light grey structure is the structure before deformation.*

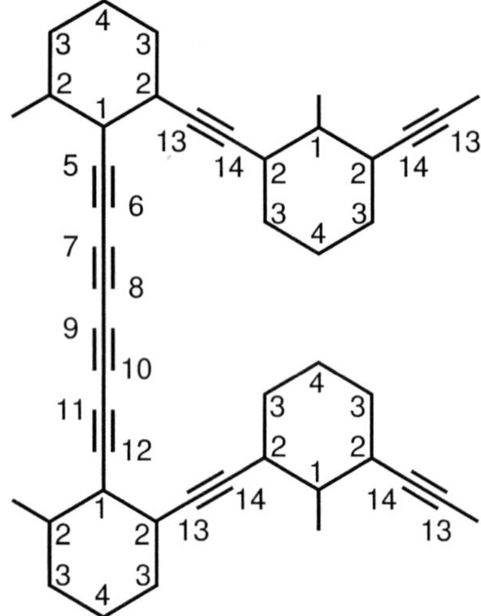

Figure 5. Atom identification numbers used in measuring the bond lengths and bond angle deformations shown in Table 2.

Table 2. Tables showing the gradients for plots of percentage strain in bond lengths and angles for loading in the Y and Z directions from 0 to 2 GPa

Bond lengths

	benzene rings			*m-chains*		
atom ID	**1-2**	**2-3**	**3-4**	**2-13**	**13-14**	**2-14**
Y	0.27	-0.01	-0.03	0.47	0.22	0.47
Z	0.17	-0.01	-0.09	-0.14	-0.06	-0.14

	n-chains								
atom ID	**1-5**	**5-6**	**6-7**	**7-8**	**8-9**	**9-10**	**10-11**	**11-12**	12-1
Y	-0.08	-0.01	-0.01	-0.01	-0.01	-0.01	-0.01	-0.01	-0.08
Z	0.50	0.26	0.27	0.26	0.26	0.26	0.27	0.26	0.50

Bond angles

	benzene rings				*m-chains*	
atom ID	**2-1-2**	**1-2-3**	**2-3-4**	**3-4-3**	**2-13-14**	**13-14-2**
Y	0.18	-0.25	0.14	0.03	-0.02	-0.02
Z	-0.54	0.24	0.13	-0.21	-0.03	-0.03

	n-chain							
atom ID	**1-5-6**	**5-6-7**	**6-7-8**	**7-8-9**	**8-9-10**	**9-10-11**	**10-11-12**	11-12-1
Y	0.00	0.00	0.00	0.00	0.00	0.00	0.00	0.00
Z	0.00	0.00	0.00	0.00	0.00	0.00	0.00	0.00

	junctions			
atom ID	**1-2-13**	**14-2-3**	**2-1-5**	**12-1-2**
Y	0.64	-0.40	-0.08	-0.08
Z	0.37	-0.62	0.26	0.28

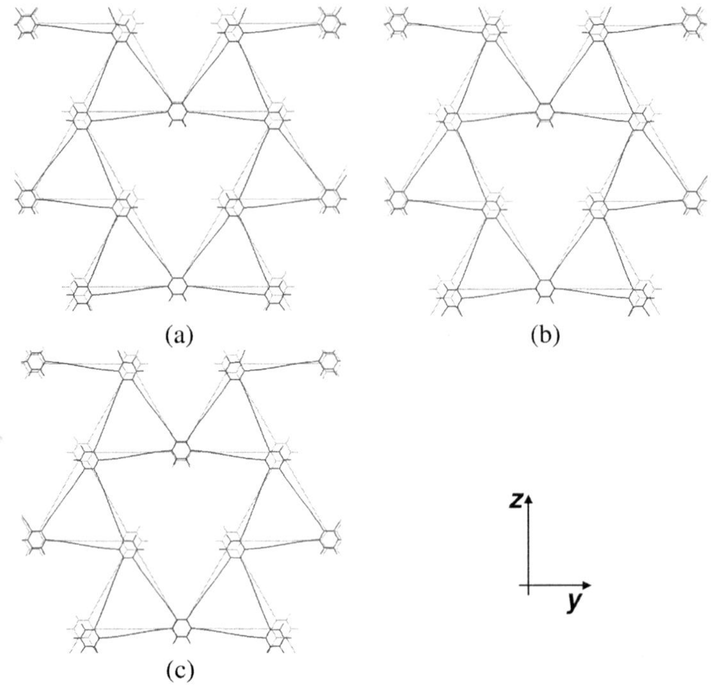

Figure 6. Deformation mechanism of the polytriangle (2B) due to a stress along (a) the Y-, (b) Z- and (c) 45° Z-axis. The light grey structure is the structure before deformation.

In the case of the calix structures, as illustrated in figures 78910, the simulations suggest that with an increase in the applied on-axis stress, *i.e.* for loading along the *Y*-axis or the *Z*-axis, the calix opens up like an umbrella, resulting in a negative Poisson's ratio. This is accompanied by an increment in the pore size of the structure, an effect which is more dominant when the number of phenyl linkages is increased. This is very significant as it confirms that these systems are indeed mimicking the behavior of the 'egg-rack' macrostructure.

Nevertheless, the 'umbrella mechanism' obtained in these simulations is not symmetrical in the sense that the opening on the 'phenyl arms' in the direction of loading is always greater than the opening of the arms in the orthogonal direction (*i.e.* there is a loss of C_{4v} symmetry which tends to become C_{2v}). One may thus refer to it as a 'defective umbrella' mechanism, which explains why the Poisson's ratio of these systems was always less negative than -1, the value of the Poisson's ratio as predicted by the idealized model of this structure (*4*). Also, it is important to note that the 'defective umbrella mechanim' is much more dominant in the smaller **3A** systems. Additionally, one should also note that in reality, the four poly-phenyl chains making up one 'umbrella' do not meet at a single point but 'around a rim' which is itself flexible. These deformations of the rims do not contribute to auxetic behavior and thus the magnitude of observed Poisson's ratio will depend on the relative contribution of the two opposing effects. For maximum auxeticity, one would require that the 'umbrella type deformations' leading to auxetic behavior are more dominant than the deformations of the rim, something which is observed more in the larger **3B** and **3C** systems when compared to the smaller **3A** systems. In this respect we note that this effect is so significant that it overturns 'imperfections' that are clearly visible in **3B** and **3C** systems when compared to **3A**. (Note: The larger **3B** and **3C** systems which are characterized by lower densities and moduli when compared to **3A**, deviate from the idealized 'egg rack' geometry as a consequence of the increased flexibility that accompanies the increase in the length of the 'polyphenyl rods'. In fact, the unstretched **3B** and **3C** networks appear as 'less ordered' when compared with the unstretched **3A** with the 'polyphenyl rods' adopting slightly bent conformations with the result that the individual calixarene building blocks do not all align with their rims being parallel to the YZ plane (the plane of the networks), thus deviating from the highly ordered macrostructure they are meant to mimic.)

If we now look at the behavior of systems when loaded at 45° to the axis, we note in the case of the **3A** system, that the umbrella mechanism is not observed, and instead the off-axis stress results in a deformation of the calixes whereby the distances between adjacent phenyl rings increases / decreases in the manner illustrated in figure 11 with the result that the calixes appear as 'sheared' and the whole structure behaves in 'wine-rack' type deformation mechanism as illustrated in the figure.

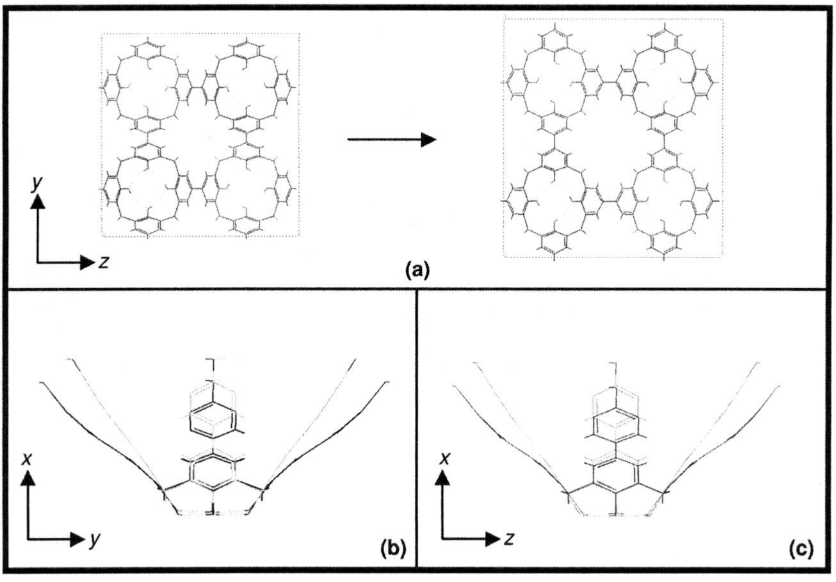

Figure 7. Deformation mechanism of the calix[4]arene (3A) due to a stress (of 20% of the y-axis Young's modulus) along the Y-axis. The light grey structure is the structure before deformation.

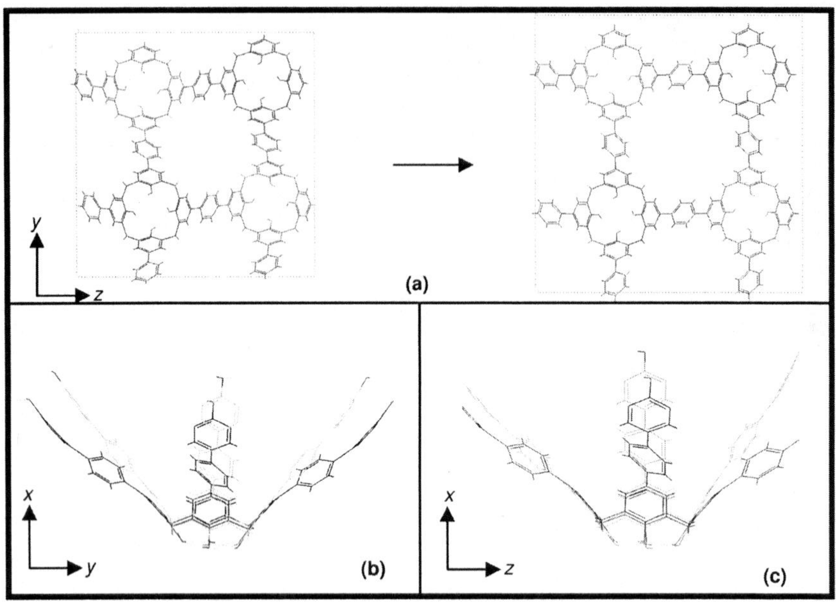

Figure 8. Deformation mechanism of the calix[4]arene (3B) due to a stress (of 20% of the y-axis Young's modulus) along the Y-axis. The light grey structure is the structure before deformation.

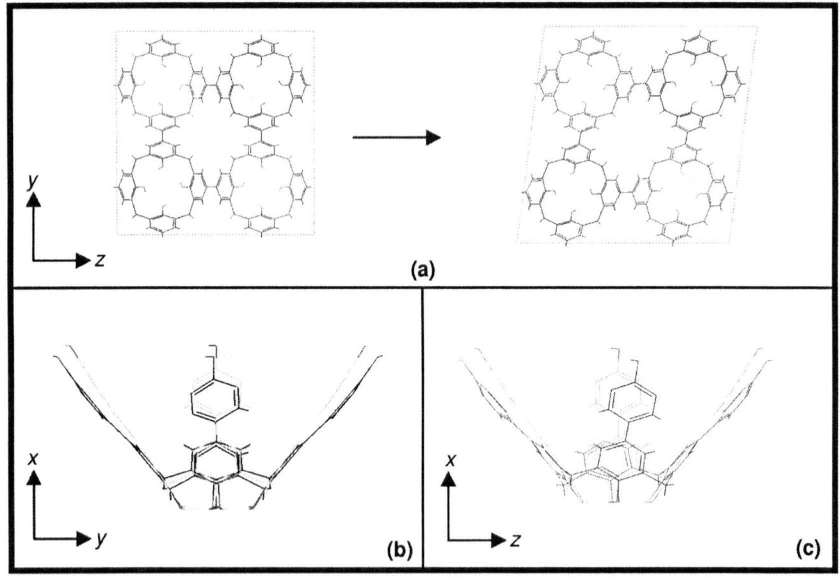

Figure 9. Deformation mechanism of the calix[4]arene (3A) due to a stress (of 20% of the 45° off-axis Young's modulus in the YZ-plane) along the 45° off-axis in the YZ-plane. The light grey structure is the structure before deformation.

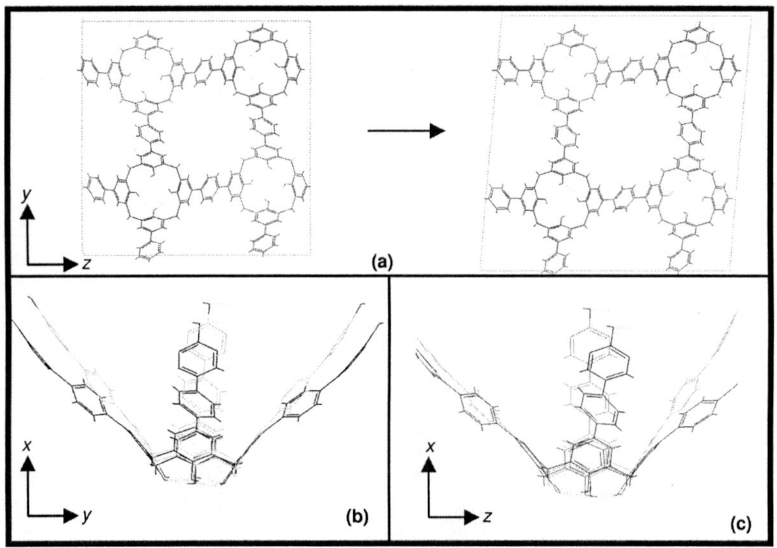

Figure 10. Deformation mechanism of the calix[4]arene (3B) due to a stress (of 20% of the 45° off-axis Young's modulus in the YZ-plane) along the 45° off-axis in the YZ-plane. The light grey structure is the structure before deformation.

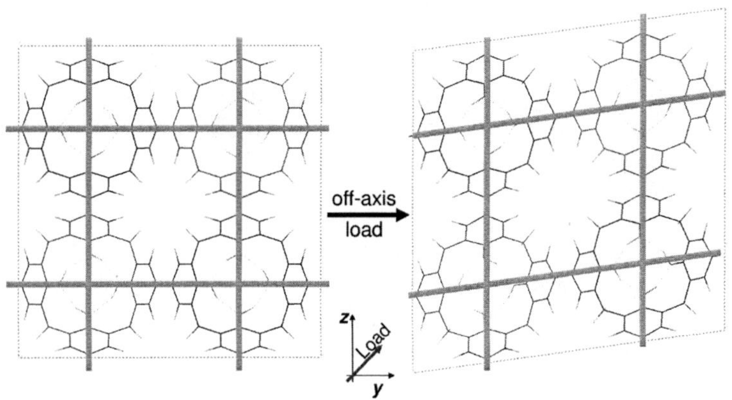

Figure 11. Schematic diagram of a 'wine rack' deformation mechanism of a calix[4]arene.

Table 3. ***The percentage changes in some dimensions as the* 3A *calyx system is stretched by 20% of the modulus in the particular direction***

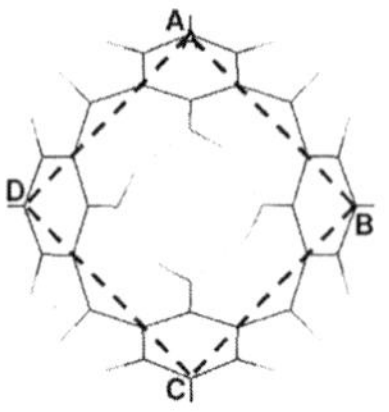

Direction of Loading	*y-axis*	*z-axis*	*45° off-axis*
AD	11.9 %	11.6 %	12.2 %
AB	11.6 %	11.9 %	-0.5 %
BC	11.9 %	11.6 %	12.3 %
CD	11.6 %	11.9 %	-0.5 %
BD	7.9 %	15.5 %	5.7 %
AC	15.5 %	7.9 %	6.4 %

Such deformations are best explained by looking at some interatomic distances associated with a typical single calix. In particular we make reference to table 3 and look at the shape of the quadrilateral ABCD where AB, BC, CD and DA represent the distances between adjacent phenyl rings in the same calyx for the **3A** system. In systems where the calixes have a 'perfect' C_{4v} symmetry, the quadrilateral ABCD would have the approximate shape of a square, a property which we observe in the undeformed conformation of **3A**. When **3A** was loaded on-axis, we observe that ABCD has an approximate shape of a rhombus where (i) both diagonal are larger than those observed in the unstretched conformation with the diagonals in the direction of the applied stress increasing more than the other diagonal (c. 16% vs. 8%); (ii) the side lengths AB, BC, CD and DA are still approximately of equal length and approximately 12% longer that those of the unstretched system. These deformations are all in line with the proposed 'defective umbrella' mechanism. Instead, when the system **3A** is loaded at 45° off-axis, a direction which is parallel to side lengths AD and BC, we note that although the diagonals increase slightly in length by *c.* 6%, an increase which suggests that the 'umbrella' mechanism is still having some contribution, we also find that whilst the lengths AD and BC increase significantly (*c.* 12%), the side lengths AB and CD remain of approximately the same length. All this clearly suggests that predominant deformation is not that of an 'umbrella' but one which is better described as a 'wine rack' mechanism (see figure 11) which has the net effect that no auxetic behavior is observed. Fortunately, this 'non auxetic' mode of deformation appears to be less predominant in the **3B** and **3C** systems with the result that these systems exhibit some degree of auxeticity even when loaded at 45° off-axis. All this suggests that the simplest of these polycalixes where the different calixes are connected directly to each other is not highly suitable for auxetic behavior and instead one should aim for ones such as **3B** or **3C**.

Conclusion

This work discussed the deformation mechanisms which result in auxetic behavior in three classes of networked polymers which have been proposed as possible candidates for auxetic materials, and, probably more importantly, the competing 'non-auxetic' mechanisms which reduce the auxetic potential of these systems. These findings are likely to be of great use to synthetic chemists not only since they may now be guided to understand the really essential features that make these systems auxetic, but also to have a clearer understanding of the strengths and limitations of these systems from a mechanical point of view.

In view of the many beneficial effects that these polymers are expected to exhibit when compared to conventional materials, we hope that this work will stimulate more experimental work aimed at synthesizing these polymers, particularly those which appear to be amenable to exhibiting auxetic behavior.

References

1. Evans, K. E.; Nkansah, M. A.; Hutchinson, I. J.; Rogers, S. C. *Nature* **1991**, *353*, 124.
2. Lakes, R. *Science* **1987**, *235*, 1038.
3. He, C. B.; Liu, P. W.; McMullan, P. J.; Griffin, A. C. *Phys. Status Solidi B* **2005**, *242*, 576.
4. He, C. B.; Liu, P. W.; Griffin, A. C. *Macromolecules* **1998**, *31*, 3145.
5. Baughman, R. H.; Galvao, D. S. *Nature* **1993**, *365*, 735.
6. Evans, K. E.; Alderson, A.; Christian, F. R. *J. Chem. Soc., Faraday Trans.* **1995**, *91*, 2671.
7. Alderson, A; Davies, P. J.; Williams, M. R.; Evans, K. E.; Alderson, K. L.; Grima, J. N. *Mol. Simul.* **2005**, *31*, 897.
8. Grima, J. N. Ph.D. Thesis, University of Exeter, 2000.
9. Grima, J. N.; Evans, K. E. *Chem. Commun.* **2000**, *16*, 1531.
10. Grima, J. N.; Williams, J. J.; Evans, K. E. *Chem. Commun.* **2005**, *32*, 4065.
11. Grima, J. N.; Williams, J. J.; Gatt, R.; Evans, K. E. *Mol. Simul.* **2005**, *31*, 907.
12. Evans, K. E.; Nkansah, M. A.; Hutchinson, I. J. *Modell. Simul. Mater. Sci. Eng.* **1994**, *2*, 337.
13. Grima, J. N.; Attard, D.; Cassar, R. N.; Farrugia, L.; Trapani, L.; Gatt, R. *Mol. Simul.* **2008**, *34*, 1149.
14. Aldred, P.; Moratti, S. C. *Mol. Simul.* **2005**, *31*, 883.
15. Wei, G. Y. *Phys. Status Solidi B* **2005**, *242*, 742.
16. Evans, K. E.; Nkansah, M. A.; Hutchinson, I. J. *Acta Metall. Mater.* **1994**, *42*, 1289.
17. Smith, C. W.; Grima, J. N.; Evans, K. E. *Acta Mater.* **2000**, *48*, 4349.
18. Grima, J. N.; Alderson, A.; Evans, K. E. *J. Phys. Soc. Jpn.* **2005**, *74*, 1341.
19. Lakes, R. S.; Elms, K. *J. Compos. Mater.* **1993**, *27*, 1193.
20. Chan, N.; Evans, K. E. *J. Cell. Plast.* **1998**, *34*, 231.
21. Choi, J. B.; Lakes, R. S. *J. Compos. Mater.* **1995**, *29*, 113.
22. Alderson, K. L.; Evans, K. E. *J. Mater. Sci.* **1993**, *28*, 4092.

23. Evans, K. E.; Caddock, B. D. *J. Phys. D: Appl. Phys.* **1989**, *22*, 1883.
24. Alderson, A.; Evans, K. E. *J. Mater. Sci.* **1995**, *30*, 3319.
25. Alderson, A.; Evans, K. E. *J. Mater. Sci.* **1997**, *32*, 2797.
26. Baughman, R. H.; Shacklette, J. M.; Zakhidov, A. A.; Stafstrom, S. *Nature* **1998**, *392*, 362.
27. Yeganeh-Haeri, A.; Weidner, D. J.; Parise, J. B. *Science* **1992**, *257*, 650.
28. Keskar, N. R.; Chelikowsky, J. R. *Nature* **1992**, *358*, 222.
29. Alderson, A.; Alderson, K. L.; Evans, K. E.; Grima, J. N.; Williams, M. *J. Metastable Nanocryst. Mater.* **2005**, *23*, 55.
30. Alderson, A.; Evans, K. E. *Phys. Chem. Miner.* **2001**, *28*, 711.
31. Alderson, A.; Alderson, K. L.; Evans, K. E.; Grima, J. N.; Willams, M. R.; Davies, P. J. *Phys. Status Solidi B* **2005**, *242*, 499.
32. Grima, J. N.; Gatt, R.; Alderson, A.; Evans, K. E. *J. Mater. Chem.* **2005**, *15*, 4003.
33. Kimizuka, H.; Ogata, S.; Shibutani, Y. *Phys. Status Solidi B* **2007**, *244*, 900.
34. Kimizuka, H.; Kaburaki, H.; Kogure, Y. *Phys. Rev. Lett.* **2000**, *84*, 5548.
35. Alderson, A.; Evans, K. E. *Phys. Rev. Lett.* **2002**, *89*, 225503.
36. Grima, J. N.; Gatt, R.; Alderson, A.; Evans, K. E. *Mater. Sci Eng., A* **2006**, *423*, 219.
37. Grima, J. N.; Jackson, R.; Alderson, A.; Evans, K. E. *Adv. Mater.* **2000**, *12*, 1912.
38. Grima, J. N.; Zammit, V.; Gatt, R.; Alderson, A.; Evans, K. E. *Phys. Status Solidi B* **2007**, *244*, 866.
39. Grima, J. N.; Gatt, R.; Zammit, V.; Williams, J. J.; Evans, K. E.; Alderson, A.; Walton, R. I. *J. Appl. Phys.* **2007**, *101*, 086102.
40. Williams, J. J.; Smith, C. W.; Evans, K. E.; Lethbridge, Z. A. D.; Walton, R. I. *Chem. Mater.* **2007**, *19*, 2423.
41. Sanchez-Valle, C.; Sinogeikin, S. V.; Lethbridge, Z. A. D.; Walton, R. I.; Smith, C. W.; Evans, K. E.; Bass, J. D. *J. Appl. Phys.* **2005**, *98*, 053508.
42. Gibson, L. J.; Ashby, M. F.; Schajer, G. S.; Robertson, C. I. *Proc. R. Soc. London, Ser. A* **1982**, *382*, 25.
43. Scarpa, F.; Bullough, W. A.; Lumley, P. *Proc. Inst. Mech. Eng., Part C* **2004**, *218*, 241.
44. Scarpa, F.; Tomlinson, G. R. *J. Appl. Mech. Eng.* **2000**, *5*, 207.
45. Wojciechowski, K. W. *Phys. Lett. A.* **1989**, *137*, 60.
46. Wojciechowski, K. W.; Branka, A. C. *Phys. Rev. A.* **1989**, *40*, 7222–7225.
47. Wood, M.; Grima, J. N.; Evans, K. E.; Alderson, A. *Xjenza* **2004**, *9*, 3.
48. Masters, I. G.; Evans, K. E. *Compos. Struct.* **1996**, *35*, 402.
49. Sun, H.; Mumby, S.; Maple, J.; Hagler, A. *J. Am. Chem. Soc* **1994**, *116*, 2978.
50. Ewald, P. P. *Ann. Phys.* **1921**, *64*, 253.
51. *Cerius2 User Manuals* (and references cited within), Accelrys, Inc.: San Diego.
52. Nye, J. F. *Physical Properties of Crystals*; Oxford University Press: Oxford, U.K., 1957.

Chapter 14

Pseudo-Blend Model of Hydrogel Immobilized Living Cells†

M. B. Plavsic,*,[1] I. Pajic-Lijakovic,[1] B. Bugarski,[1] J. Budinski-Simendic,[2] V. Nedovic,[3] and P. Putanov[4]

[1]Faculty of Technology and Metallurgy, Belgrade, Serbia
[2]Faculty of Technology, Novi Sad, Serbia
[3]Faculty of Agriculture, Belgrade, Serbia
[4]Serbian Academy of Sciences and Arts, Belgrade, Serbia
***plavsic@tmf.bg.ac.rs**

†Dedicated to Professor F. E. Karasz on the occasion of his 75th birthday

Living cell organization in clusters is considered in terms of cell-cells and cells-microenvironment interactions as important factor in fundamental understanding of cell behavior in formation and grow of living tissues, wound heeling and development some serious illness, but also in present tissue engineering, immobilization of cells in polymer matrices and culture grow. Significant importance of parameters as compacivity and compressibility for quantification of cell behavior is found as well as possibility for modeling cell organization in terms of classical models of polymer structure as blends. The theory of Karasz and coworkers on miscibility widows in polymer blends is used for interpretation of some polymer-cell relations.

It is well known that cells are morphological unit structures of all living organisms (*1*) Some of them can function as autonomous living entities as well. Looking broader and deeper than morphological studies, one can say that the fine border line between living and nonliving world is just a part of the cell. Naturally, all fundamental sciences have exposed their interest for cells, especially physics, from the very beginnings of modern physics and development devices for exact measurements, till present studies of quantum effects, especially in integrative biophysics and quantum medicine (*2*). But modern high technologies as tissue engineering, pharmaceutical engineering and bioengineering (especially with production of bio-active substances in bioreactors) are also looking in parallel with detail biochemical information, for global properties (in de Gennes' terminology (*3*)) of cell systems. But, the main parts of cell structure build macromolecules. Also in tissues, the next main unit in bio-organization hierarchy, the dominant role again in structure formation belongs to polymer fibers and networks. It follows quite naturally that polymer science should expose its interest for cells and to do it from both, fundamental and applied aspects.

Blends are today one of the most important fields in both, polymer science and engineering. Accumulated data and experience, methods and techniques, as well as theoretical models provide powerful tools to be applied also in the border and interdisciplinary areas of modern polymer practice. For the large part of that achievements polymer scientist are debt to Professor F. Karasz and his coworkers. Moreover, their theory on cross combinations of interaction parameters or blend "miscibility windows" (*4*, *5*),can be also applied to a number of meta -stable systems in new areas of research in particular for bio-systems..

It is well recognized today that, although living systems obviously obey the lows well known in physics and chemistry, some additional parameters are required for efficiently modeling their functions. Those parameters are also a kind of "windows" for fine tuning structure of living world. Just there, different scientific fields again combine and "mix" looking for some cross-sections of knowledge to understand new qualities. One of that is a new science of complexity, at the cross- sections of modern physics, chemistry and biology (*6*).

Statistical Mechanics and Cell Systems

T.L. Hill raised among the first the issue of applicability of statistical mechanics methods to small systems (*7*). Macroscopic thermodynamic functions and mathematical interrelations between these functions are defined for macroscopic, strictly saying, infinite system only. But if we wish to pursue thermodynamic tools on a smaller scale, we are faced earlier or later with the problem of single particle thermodynamics. It can be macromolecule, colloidal particle, liquid droplet or even the cell. The answer is of course simple, although not so obvious, similarly to other deep wisdoms of thermodynamics. We will return to it later. The issue is mentioned here because some new publications today say that statistical mechanics of cell does not exist. Cell as the system is far from thermodynamical equilibrium and particles inside vary in number from hundreds to just a few of them of the same kind! Still, J. von Neumann, suggested long time ago the formulation of mathematical model later called "cellular automata" relating cell behavior with statistical mechanics (*8*). Very interesting results till now are obtained about integrated action of large number of cells in an ensemble. But here, should be noted also that large number of related problems to cell behavior still stay unsolved.

The cellular automata models consist of discrete agents or particles, which occupy some or all sites of a regular lattice (*9*). Originally, a cell was represented simply as physical point with one or two properties, and later as physical body with fixed position in lattice. In the third model generation, both static and dynamic algorithms describing cells transitions are used. The particles have one or more internal state variables, which may be discrete, or continuous with a set of rules describing the evolution of their state and position. Both, the movement and change of state of particles depend on the current state of the particle and those of neighboring particles. Again these rules may either be discrete or continuous what can be formulated by ordinary differential equations, deterministic or probabilistic. The evolution rules can be applied in steps, e.g. a motion or transport step followed by a state change or interaction step. Updating can be synchronous or stochastic. At one extreme the rules may approximate well known continuous partial differential equations, at the other may resemble the discrete logical interactions, using Boolean algebra. The models are applied to describe cell-cell and cell-environment interactions mainly using phenomenological local rules. Till present are considered simulation possibilities for a broad range of biological examples ranging from bacteria and slime model amoebae, to chicken embryonic tissues and tumors in humans. Moreover, the method is extended to a number of fields of quite different character than cell systems, e.g. economy and computer science, with remarkable success. It participates in development of computer sciences and huge grow of computer technology what synergistically supported development of the method itself. (Synergism will be the word we have to use very often in the following text as well.) But, beyond it, some essential starting targets are still not jet achieved. It has to be discussed more in detail.

We can describe today surprisingly complicated structures combining simple states (or "decisions") of constitutional units with help of computers. Simple "on/off presentation" of those states is convenient for computer work itself and such model programming. The large number of such unites can be used as presentation of an ensemble of particles, or ensembles of states for one particle. In both cases we deal again with methods of "classical" Statistical Mechanics. But how cells make those simple decisions or realize their simple states? They are too large to perform thermal movements as particles in classical thermodynamic systems. But, if they respond deterministically, e.g. as moving macro objects, than for the same influence from outside, they will be all in the same state. Some statistical mechanics must be involved to their functions.

Also some processes involved here can not be reduced to elementary steps. It is the feature of complex systems. It seems that cells follow that behavior, at lest in some parts of their action. The idea is that if higher level properties really do emerge in complex systems, yielding wholes that are more than the sum of the parts, than explanations of these systems must refer to the higher level properties. Everything cannot be reduced to behavior descriptions of lower level parts.

P. Bak (*10*) illustrated the above principle of non-reductionism with sand pile model. When the growing sandpile reaches the state where it is subject to catastrophic collapse the pile itself is a functional unit, not the single grains of sand. No reductionist approach makes sense. To predict a catastrophic avalanche in traditional reductionist terms one would have to measure everything everywhere, with absolute accuracy, which is impossible. Then one would have to perform an accurate computation based on this information, which is equally impossible.

The sendpile is here just visualization for one very abstract problem. Some theoretician in the science of complexity as J. Holland (*11*) insist on abstraction in the sense of sharp –edged, unambiguous, precise terms ridded as far as possible of qualitative or phenomenological concept, or say: better even use numbers than terms. But, S. Edwards has made one unusual proposal for one new term that is at the same time very abstract but also easy to visualize (*12*). It deals with entropy as "higher level property" as quoted above, but at the same time (and in a kind of astonishing way, but comfortable too,) makes equality between free energy and volumes of the cluster. Of course the complexity of cluster states is the silent aspect of relation between changes of its free energy and volumes. It is literally related to real sendpile, but also to some new kinds of entropy, recognized in statistical mechanics of jammed state of matter (*13–15*).

Briefly saying, Edwards and coworkers have considered the volume of a cluster as the main parameter describing jammed state of particles (*16*, *17*). As the model system they proposed a cluster of hard irregular shape grains with interaction forces (e.g. friction) between them preventing flow, but exposed to tapping forces from environment and changing volume in that way, e.g. a pile of sand. At the first site it is quite different system from cell systems considered here. But, more careful examination finds many of their properties that can be defined in the same terms. They introduced a new and very significant parameter for description of grain clusters named compactivity X, representing the change of cluster volume V with entropy S for given number of particles N (i.e. considering it as a canonical ensemble):

$$X = \left(\frac{\partial V}{\partial S}\right)_N \qquad (1)$$

Probability distribution for cluster states is than:

$$P = e^{(Y-W)/\lambda_E X} \qquad (2)$$

where Y is effective volume of the cluster (corresponding to the free energy F in classical statistical mechanics) and W is the "volume function" corresponding to the Hamiltonian and λ_E is constant adjusting dimensions. Than, we can write

$$Y = V + X\frac{\partial Y}{\partial X} = V - XQ \qquad (3)$$

The partition function relating Y with the volume function of a particle w_{gr} can be written in the form:

$$e^{\frac{-Y}{\lambda_E X}} = \int \Omega_w e^{\frac{-w_{gr}}{\lambda_E X}} d[all] \qquad (4)$$

where integration goes over all positions of all particles in the cluster. For estimation of possible values of w_{gr} can be used formula:

$$w_{gr} = v_0 + (v_{max} - v_0)\xi^2 \tag{5}$$

where v_0 is the grain volume, v_{max} is maximal volume the grain can occupy in a cluster and ξ represents its degree of freedom to adjust its position and orientation in the cluster. It can have values in the range 0 to 1. For loose packing of particles, we can take $w_{gr} = w_p$ and a grain has high degree of freedom in such system so ξ values are close to 1, giving w_p values close to v_{max}. For dense packing ξ is close to 0 giving w_p values close to v_0. Parameter ξ provides some possibilities to account orientation effects for grains of anisotropic shape in cluster, using eq4.

The copactivity relates the volume of particle cluster to cluster entropy. It is a property of the cluster as a whole, giving possibilities that two other important characteristics can be recognized: emergency and holism.

We talk about emergency here in the sense that if the whole is greater than the sum of its parts then somewhere along the way from parts to whole, something in addition to the parts must have "emerged". In the case of sendpile, the necessary information is distributed through the whole of the pile. It is a matter of the interlinked balances of force upon every grain in the pile, the shape of every grain and so on. Therefore understanding must proceed on a holistic basis.

The whole is greater than the sum of its particles, as pointed Kauffman (*6*). It is the main formulation in "holism". It follows that understanding how interact parts of a biological system, e.g. "modules" inside a cell, polymer network structures in it and genes is just as important to understanding as the cell "classical" molecular interactions (say in biochemical approach). It is a realization that is beginning to spread. The cell as complex system has been taken as one whose properties are not fully explained by an understanding of its component parts.

But what about structures that exist in -between the unit structures and the whole? One can say that something like this dos not exist in model system like sendpile. But the main challenge where failed cellular automata model is embrio morphogenesis. Embryos and tissue seem to obey rules differing from the physical rules been associated with differential equations of position and state changes in this model: their forms seem to result from expression of intrinsic highly complex genetic programs. But, again surprisingly, embryos, organs and healing and regenerating tissues assume many forms resembling (basic and simple) behavior of those physic produces in non-living matter as crystals. More precisely, resemble processes of polymer crystal grow and melting. It is also close to another basic mechanism in polymer science: phase transitions and order parameter changes in polymer blends.

Proposing this relations between cell and blend organization we also try to complete one "whole of ideas" from morphology facts (mentioned at the very start) till abstraction of collective behavior in terms of statistical mechanics of fields. We also believe that methods developed for energy interactions in super -molecular polymer structures can be useful for both, and make connections between inter- and intra- cell interactions. The cross terms of these interactions, are of especial importance. It will be considered in the next section.

Supermolecular Polymer Structures and Miscibility Window Paradigm

In his classical work on thermodynamics of polymer solutions P.J. Flory (*18*) proposed simple but very useful equation for free energy of mixing, that has been applied later also to polymer blends:

$$\Delta G_M = \Delta H_M - T\Delta S_M = RT(n_1 \ln \varphi_1 + n_2 \ln \varphi_2) + BV_1 n_1 \varphi_2 \quad (6)$$

where n_i is number and φ_i volume fraction of component i, R is the gas constant, T temperature and B the interaction energy density. If B is not function of temperature eq6 can be used to predict upper critical solution temperature behavior (UCST), since entropic contributions become more favorable at higher temperatures whereas the enthalpic contribution is constant. However if there is an entropic contribution to interaction parameter

$$B = B_H - TB_s \quad (7)$$

the lower critical solution temperature (LCST) can be obtained. The origins of phase separation on heating can be attributed to tree possible causes, more than one of which may be operative (*18–22*):

1. Contributions from the volume change ΔV_M on mixing to the free energy can become more unfavorable at higher temperatures.
2. The heat of mixing ΔH_M may be temperature dependent. This may especially be true if miscibility rises due to specific interaction which may dissociate on heating.
3. There may be unfavorable entropy contributions in ΔS_M which could rise from non random mixing, for example associated with specific interactions. This may not necessarily result in a temperature dependence consistent with eq7.

It has been noted that copolymers are more likely to be miscible with other polymers than homopolymers are. Addition of attracting groups to two basically different chains can change properties of the system as a whole. Now are well recognized interesting cases with so called miscibility windows range of copolymer composition over which it is miscible with another polymer (*4*, *5*, *19*) Of especial interest is combination of a copolymer with some homopolymer. For example some ethylene/vinyl acetate copolymers are miscible with PVC where as neither poly(ethylene) nor poly(vinyl acetate) are miscible. It is explained with contribution of cross terms i.e. unfavorable interactions between different segments in the same polymers (*4*, *5*, *19*). Such basic terms have been also considered describing the solution properties of copolymers. As suggested a copolymer is more likely to be miscible with other polymer since mixing can reduce the number of unfavorable contacts between the two types of segments. The process can be simply represented as: the (1) and (2) units of the copolymer mix with the unit (3) of the other polymer. If the interaction energies between the various segments are B_{12}, B_{13}, and B_{23} than the interaction energy for the blend will be given by:

$$B_{blend} = B_{13}\varphi_1' + B_{23}\varphi_2' - B_{12}\varphi_1'\varphi_2' \tag{8}$$

Thus if B_{12} is positive and large enough to counteract B_{13} and B_{23} the overall interaction term can be negative. But, F. Karasz and coworkers (*4, 5*) pointed that also LCST behavior in polymer blends can be result of compressible nature of the system and directional specific character of their intermolecular interactions or combination of both. For example PVC is miscible with PCL in the broad range of temperatures and high molecular weights but, with PMMA for not high molecular weight and just some temperature. It can be explained with larger los in entropy for formation interactions between PMMA and PVC than for PVC with PCL. The carbonyl group in poly(caprolactone) is in the main chain but in poly(methylmetacrilatr) it is the side group. The orientational effects can appear in the course of cell cluster formation on macromolecular fibers in tissue or polymer matrix as the cell support in general (*22–24*). This analogy will be elaborated in the next section. Also analogous to eq6, and in accordance to analyses presented in previous section we can write for a cell system

$$\Delta Y_M = \Delta V_M - X\Delta Q_M \tag{9}$$

where effective volume of the cell cluster can depend on several additional factors in particular metabolite components producing the third kind of interactions in cell culture systems, and negative electrical charges of cells that influence not only cells but also the matrix. But modern theories of electro- viscoelasticity provide background for further understanding of clustering effects in cell colonies (*25–27*). It seems to be forgotten in such kind of simplified model analyses of complex cell system states, interacting with polymer matrix groups, but can be of significant influence to the pressure in the system (*27–55*). It follows in general that pressure effects are of especial importance in "phase relations" of cell-matrix systems, what will be considered in next section more in detail.

Mathematical Formulation of the Model

We will start considering of the cell–polymer matrix pseudo-blend structures by using an alginate micro-bead as typical example. It is produced by dropping homogenized polymer/cell mixture into calcium chloride solution. Calcium ions penetrate into droplets and make ionic junctions of alginate polymer network. In that way cells at primordial colony level were randomly spread over the volume of the bead, dominantly single. They can move and join each other (in the neighborhood) to small groups that further develop to clusters but are prevented to join clusters or to escape out of the bead, by the polymer network. The dynamics of structural changes within the pseudo-blend is regulated by stress difference between cell population and polymer matrix. Generation of internal stress within cell population is caused by cell motion and growth, while generation of external stress within hydrogel matrix is the resistance to cell actions. So, the stress within matrix has the feedback loading action, as compression of cell population.

For further interpretation of such complex phenomenon, we will introduce two time scale, i.e. long, growing-time t and short, loading-time τ. Loading-time is close to the time for relaxation of both subsystems. But, time scales of t and τ (growing and loading times), is different many orders of magnitude.

For further interpretation it is necessary to formulate the changes of configurational Helmholtz free energy as the sum of two contributions:

$$dF(\tau,t)=dF_m(\tau,t)+dF_C(\tau,t) \tag{10}$$

where $dF_m(\tau,t)$ is the matrix subsystem contribution, while $dF_C(\tau,t)$ is the cell subsystem contribution. The change of Helmholtz free energy for cell subsystem is expressed as:

$$dF_C(\tau,t)=\frac{\partial F_C(\tau,t)}{\partial N(t)}dN(t)+\frac{\partial F_C(\tau,t)}{\partial V_C(\tau,t)}dV_C(\tau,t) \tag{11}$$

where $V_C(\tau,t)$ is the volume of the cell subsystem which represents the sum of the volume of clusters and $\frac{\partial F_C(\tau,t)}{\partial N(t)}=\mu_{eff}(t)$ is the effective chemical potential where $N(t)$ is the number of cells per micro-bead. Now we can write $\frac{\partial F(\tau,t)}{\partial V_C(\tau,t)}=-\Delta P(\tau,t)$ for the stress generated in the system. The optimal number of cells per microbead is realized by the action of the difference between internal pressure generated within clusters and external pressure in polymer hydrogel matrix, The pressure difference will be expressed as:

$$\Delta P(\tau,t)=\Pi(\tau,t)-P_m(\tau,t) \tag{12}$$

where $P_m(\tau,t)$ is in fact the external stress within polymer matrix, while $\Pi(\tau,t)$ is the internal stress generated within the cell subsystem. We only consider the pressure for simplicity, but we could use the complete stress tensor and write $\Delta P_{ij}(\tau,t)=\Pi_{ij}(\tau,t)-P_{ext\ ij}(\tau,t)$. For interpretation of the internal stress, we will use the virial stress (*30*, *31*). The virial stress in general has two contributions from particles, reflecting their kinetic and potential energy. However, dependently of the phenomenon of interest, some authors have considered only potential contribution. Edwards (*16*, *17*, *31*) used virial stress to formulate the dynamics within sandpile clusters by considering only the potential contribution. In our case cells are particles having also kinetic energy. On that base, it will be expressed as:

$$\Pi(\tau,t)=\frac{1}{V_C(\tau,t)}\left[\sum_{i=1}^{N(t)} m_i v_i v_i + \sum_{i=1}^{N(t)} \sum_{j=1}^{\vartheta(t)} r_{ij} \frac{\partial \phi_{ij}}{\partial r_i}\right] \tag{13}$$

where m_i is mass of single cells and v_i is cell velocity, $\vartheta(t)$ is the number of nearest neighbors in contact with cell *i*. Cells motion has been interpreted as the motion of Brownian particles but produced by its own energy (*1*). The force on cell *i* due to interaction with cell *j* is $F_{ij}=\frac{\partial \phi_{ij}}{\partial r_i}$, while r_{ij} is the distance vector between cells *i* and *j*.

The virial stress $\Pi(\tau,t)$ depends on: (1) the volume of cell subsystem, (2) the number of cells per microbead and (3) the motion of cells within clusters. Accordingly, virial stress increases with the number cells and decreases with expansion of the volume of cell subsystem. Expansion of cell population placed within the clusters occurs during the many relaxation cycles caused by the loading conditions. Every relaxation cycle of cell population produces the feed-back loading to the surrounding polymer hydrogel. The phenomenon has the oscillatory nature.

The external stress generated within hydrogel matrix can be expressed as:

$$P_m(\tau,t)=P_0(t)+\Delta P_m(t)\ e^{-\frac{\tau}{\tau_R}} \tag{14}$$

where first term of eq14 i.e. $P_0(t)$ represents the total averaged stress within matrix for corresponding number of cells $N(t)$, while the second term represents kinetic fluctuation contributions due to permanent motion of cells within clusters and their pressing upon the surrounding. Clusters expansion causes the local compression of surrounding matrix. Such compression further causes the stress increase for the averaged value $\Delta P_m(t)$. The second term of eq14 describes also the relaxation of surrounding polymer matrix, while τ_R represents the relaxation time. After cluster expansion at $\tau=0$, external stress reaches its maximal value, i.e. $P_m(0,t)=P_0(t)+\Delta P_m(t)$. The matrix is excited. After relaxation cycle, for $\tau \rightarrow \tau_{eq}$, the perturbation term will be equal and stress in matrix reaches the minimal value. Every cluster expansion produces the relaxation cycle of surrounding polymer medium (*19–22*). Accordingly with the fact that many relaxation cycles occurred during the time t, the suitable expression of stress difference previously formulated by eq12 can be given as periodic functional form:

$$\Delta P(\tau,t)=P_0(t)-A(t)\cos(\omega\ \tau) \tag{15}$$

where $\omega = \frac{2\pi}{\tau_R}$ and $A(t)$ is amplitude of stress which depends on the number of cells per micro-bead. In that way the interpretation of the phenomenon through effective chemical potential μ_{eff} needs nonlinear functional form: $\mu_{eff}(t) = \beta_1 N(t) - \beta_2 N(t)^2$ (β_1 and β_2 are the constants). It is $\frac{d\mu_{eff}(t)}{dN(t)} /_{N=N_{op}} = 0$ for the optimal number of cells per microbead N_{op}. The optimal number of cells per microbead is realized by the action of the difference between external stress generated within surrounding polymer hydrogel and internal stress generated within clusters of cells. From the condition of the minimum of effective chemical potential, we can determine the constant β_1 as $\beta_1 = 2N_{eq}\beta_2$.

Some critical changes of the number of cells per microbead could be treated thermodynamicaly as pseudo-phase transformation (*32*, *33*). On that base, we have used as the modification of thermodynamical models on mesoscopic level introducing the particular form of free energy functional (*36–40*). Such modifications connect the thermodynamical formulation on one side with the

$$\frac{dN(t)}{dt} = \Gamma\, \mu_{eff}(t) \tag{16}$$

where Γ is constant. Accordingly, the number of cells per microbead is $N(t) = \frac{K \cdot N_0 e^{\beta_1^{\bullet} t}}{K + N_0(e^{\beta_1^{*} t} - 1)}$ (where model parameters are $\beta_1^{*} = \Gamma\beta_1$ and $K = \beta_1 / \beta_2$).

The changes of the volume of cell subsystem can be expressed as:

$$dV_C(\tau,t) = \frac{\partial V_C(\tau,t)}{\partial\, N(t)} d\, N(t) + \frac{\partial V_C(\tau,t)}{\partial \Delta P(\tau,t)} d\, \Delta P(\tau,t) + \frac{\partial V_C(\tau,t)}{\partial S_C(t)} d\, S_C(t) \tag{17}$$

where $S_C(t)$ is the configurational entropy of cell subsystem and $\frac{\partial V_C(\tau,t)}{\partial N(t)} = \bar{V}_C(t)$ is the specific volume of single cell, $\frac{\partial V_C(\tau,t)}{\partial \Delta P(\tau,t)} = -\beta_{TC} V_C(\tau,t)$, β_{TC} is compressibility of cell subsystem at constant temperature, $\frac{\partial V_C(\tau,t)}{\partial S_C(t)} = X_C$ is compactivity of cell subsystem. By the analogy with the rearrangement of powders suggested by Edwards, by eq5 we will connect the volume changes with the degrees of freedom ξ. We can introduce for orientation of cell two degrees of freedom ξ_1 and ξ_2. When $\xi = 0$, the cells are "well oriented" and specific volume of a cell $\frac{\partial V_C(\tau,t)}{\partial N(t)}$ is equal to its minimal value, i.e. $\bar{V}_{C\min} = \langle v \rangle$ (where $\langle v \rangle$ is averaged volume of single cell). It means that free volume is minimal. When $\xi = 1$, specific volume per cell $\frac{\partial V_C(\tau,t)}{\partial N(t)}$ is equal to its maximal value, i.e. $\bar{V}_{C\max}$ and free volume is maximal. We can also interpret the process of compaction of cell population as Ornstein-Uhlenbeck process for the degrees of freedom ξ_i, $i = 1,2$. It is in accordance with the fact that all cells are driven stochastically in a bead in bioreactor (*37*, *38*), by random force with zero correlation time (*36–45*). Therefore, we could describe the phenomenon using Langevin equation in the form:

$$\frac{d\xi_i(t)}{dt} = -\frac{1}{\gamma}\frac{\partial W(\xi(t))}{\partial \xi_i(t)} + f_i(t) \tag{18}$$

where $f_i(t)$ is stochastic random force which is formulated as white noise with correlation function $\langle f_i(t) f_i(t') \rangle = 2\delta_{ij}\delta(t-t')$, while γ characterizes the frictional resistance imposed on the cell by its nearest neighbors. As Edwards has proposed (*17*) quantity $W(\xi_i(t))$ represents the analogy of Hamiltonian. Than

$$\xi_i(t) = \xi_i(0)e^{-\gamma t} + \int_0^t e^{-\gamma(t-t')} f(t')dt' \tag{19}$$

Averaging over the ensemble of cells we get:

$$\langle \xi_i(t) \rangle = \xi_i(0)e^{-\gamma t} \tag{20}$$

where the initial value for ξ_i is $\xi_i(0) = 1$. At $t \to t_{eq}$ the degree of freedom tends to zero, i.e. $\xi_i \to 0$. Accordingly, the Fokker-Planck equation is suitable to formulate the density of cell distribution with loading-time τ- with changes of degree of freedom ξ_i as:

$$\frac{\partial \rho_C(\xi_i,\tau)}{\partial \tau} = -\frac{\partial}{\partial \xi_i}\left(\frac{1}{\gamma}\frac{\partial W(\xi_i)}{\partial \xi_i}\ \rho_C(\xi_i,\tau)\right) + D_i \frac{\partial^2}{\partial {\xi_i}^2}\rho_C(\xi_i,\tau) \tag{21}$$

Finally, we can formulate in the first approximation the distribution density as the function of degrees of freedom ξ_i with time τ-:

$$\rho_C(\xi_i,\tau)=\frac{1}{\sqrt{2\pi\sigma_i^{2}(\tau)}}e^{-\frac{\left(\xi_i-\langle\xi_i(\tau)\rangle\right)^2}{2\sigma_i^{2}(\tau)}} \tag{22}$$

where $\sigma_i^{2}(\tau)$ is the variance which has the meaning of relaxation rate of degrees of freedom. The entropy of cell subsystem is in first approximation related to cell orientations:

$$S_C(t,\tau)=-k_B\, ln\prod_{i=1}^{2}\rho_C(\xi_i,t,\tau) \tag{23}$$

where k_B is Boltzmann constant and $\rho_C(\xi_i,\tau)$ is meta -stable distribution density of cells that is related to entropy contributions to cluster reorganizations in the sense of conclusion from the previous section.

Pressure effects and dynamics of cell organization, proposed here to be described and quantified in particular by compactivity of polymer clusters and compressibility can be related to number of vital functions but also to serious illness as cancer (*40–55*).

Conclusions

Problems of description and quantification of living cell organization in clusters is considered in terms of modern statistical mechanics methods. The problem of cell compaction especially with orientation, is elucidated in several aspects using statistical mechanics of jammed state, statistical mechanics of fields and irreversible processes. Physical parameters such as energy, temperature and compressibility are considered in parallel with new one from statistical mechanics of jammed matter: the compactivity. Interactions forces between cells and polymer matrix either polymer fibers and networks in extracellular tissue matrix, or scaffolds in tissue engineering, different cell colony supports, or immobilizing matrix of polymer beads, are found to follow similar pattern in interaction potentials and orientational entropy effect as described in theory of blend miscibility windows of F. Karasz and coworkers.

In that sense are considered possible applications of the compastivity, for living cell systems It can be understood as the number of ways it is possible to arrange the grains in the system with increase of volume ΔV such that disorder is ΔS. The compactivity is proposed by S. Edwards for systems of hard particle with friction. Although soft, cells have well defined volume and inter-cell interactions in analogy to friction forces between grains. But, cells are complex systems. Their complexity arises in two ways: first as an emergent property of the interaction of a large number of autonomously motile cells that can self-organize.

Cells need not firm thermodynamically equilibrated structures. As the second cells have a complex feed -back interaction with their environment. Cells can modify their surroundings by e.g. secreting diffusible or non-diffusible chemicals. Their environment in turn causes changes in cell properties e.g. differentiation by changing the levels of gene expression within the cell. Still same analogy between cell colonies with polymer blends as well as crystals is evident and can be used for efficient mathematical treatment in modeling cell cluster growth for practical and fundamental purpose. In that sense new model of pseudo-blend organization of cells in clusters is proposed.

References

1. Murray J. D. *Mathematical Biology*; Springer: Berlin, 1988.
2. Rakovic D. *Integrative Biophysics Quantum Medicine and Quantum Holographic Informatics*; IASC-IEFPP: Belgrade, 2009.
3. De Gennes, P. J. *Scaling Concept in Polymer Physics*: Cornell University: Ithaca, NY, 1979.
4. ten Brinke, G.; Karasz, F. E; McKnight, W. S. *Macromolecules* **1983**, *16*, 1827.
5. ten Brinke, G.; Karasz, F. E *Macromolecules* **1984**, *17*, 815.
6. Kauffman, S. *At Home in the Universe: The Search for the Laws of Self-Organization and Complexity*; Oxford University Press: Oxford, 1985.
7. Hill, T. L. *Thermodynamics of Small Systems*; Benjamin, Inc.: New York, 1963.
8. von Neumann, J. *Collected Works*; Taub, A. H., Ed.; Pergamon Press: Oxford, 1961−1963.
9. Wolfram, S. *Rev. Mod. Phys.* **1983**, *55*, 601.
10. Buk, P. *How Nature Works: The Science of Self-Organization and Criticality*; Springer: New York, 1963.
11. Holland, J. H. *Emergence: From Chaos to Order*; Addison-Wesley: Reading, MA, 1998.
12. Edwards, S. F. *J. Stat. Phys.* **2004**, *116*, 29.
13. Makse, H. A.; Kuchan, J. *Nature* **2002**, *415*, 614.
14. Behringer, R. *Nature* **2002**, *415*, 594.
15. Song, C.; Wang, P.; Makse, H. *Nature* **2008**, *453*, 629.
16. Edwards, S. F. *Phys. A* **2005**, *353*, 114.
17. Edwards, S. F. *J. Phys. A: Math. Theor.* **2008**, *41*, 3240.
18. Flory, P. J. *Principles of Polymer Science*; Cornel University Press: New York, 1971.
19. MacKnight, W. J.; Karasz, F. E.; Freed, J. R. Phase and Relaxation Transitions in Solid Polymer Mixtures. In *Polymer Blends*; Paul, D. R., Newman, S., Eds.; Academic Press: New York, 1978; Chapter 5.
20. Plavsic, M. B.; Pajic-Lijakovic, I.; Putanov, P. Chain Conformational Statistics and Mechanical Properties of Elastomer Blends. In *New Polymeric Materials*; Korugic-Karasz, L., MacKnight, W., Martucelli, E., Eds.; ACS

Symposium Series 916; American Chemical Society, Washington, DC, 2005; Chapter 19.

21. Plavsic, M. B. *Materials Science and Engineering of Polymeric Materials*; Sci. Book: Belgrade, 1996.
22. Plavsic, M. B.; Pajic-Lijakovic, I.; Bugarski, B.; Putanov, P. *Mater. Manuf. Processes* **2009**, *24*, 1.
23. Plavsic, M. B. *Mater. Sci. Forum* **1996**, *214*, 123.
24. Plavsic, M. B.; Pajic-Lijakovic, I.; Plavsic, M. M. *Int. J. Mod. Phys. B* **2010**, *24*, 813.
25. Spasic, A. M.; Hsu, J. P. *Finaly Dispersed Particles*; CRC, Taylor and Francis: New York, 2006; Volume 15, Chapter 1, p 26.
26. Spasic, A. M. *Chem. Eng. Sci.* **1992**, *47*, 3949.
27. Gennaro, A. M.; Luquita, A.; Rasia, M. *Biophys. J.* **1996**, *71*, 389.
28. Clausius, R. J. E. *Philos. Mag.* **1870**, *40*, 127.
29. Kirkwood, J. G. *J. Chem. Phys.* **1946**, *14*, 180.
30. Marc, G.; Millan, W. G. *Adv. Chem. Phys.* **1985**, *58*, 209.
31. Goldbert, P.; Goldenfeld, N.; Sherrington, D. *Steeling the Gold: A Celebration of the Pioneering Physics of Sam Edwards*; Oxford Science Publications, Oxford, 2004.
32. Ala-Nissila, T.; Majaniemi, S.; Elder, K. *Lect. Notes Phys.* **2004**, *640*, 357.
33. Landau, L. D.; Lifshitz, E. M. *Statistical Mechanics*; Pergamon Press: Oxford, 1984.
34. Pajic-Lijakovic, I.; Plavsic, M.; Nedovic, V.; Bugarski, B. *J. Microencapsulation* **2007**, *24*, 420.
35. Pajic-Lijakovic, I.; Bugarski, D.; Plavsic, M.; Bugarski, B. *Process Biochem.* **2007**, *42*, 167.
36. Pajic-Lijakovic, I.; Plavsic, M.; Bugarski, B.; Nedovic, V. *J. Biotechnol.* **2007**, *129*, 446.
37. Pajic-Lijakovic, I.; Plavsic, M.; Nedovic, V.; Bugarski, B. *Minerva Biotechnol.* **2008**, *20*, 99.
38. Pajic-Lijakovic, I.; Bugarski, B.; Nedovic, V.; Plavsic, M. *Minerva Biotechnol.* **2005**, *17*, 245.
39. Pajic-Lijakovic, I.; Ilic, V.; Bugarski, B.; Plavsic, M. *Eur. Biophys. J.* **2009**, DOI: 10.1007/S00249-009-0554-6.
40. Peppas, N. A.; Huang, Y.; Torres-Lugo, M.; Ward, J. H.; Zhang, J. *Annu. Rev. Biomed. Eng.* **2000**, *2*, 9–29.
41. Reif, M.; Fernandez, J. M.; Gaub, H. E. *Phys. Rev. Let.* **1998**, *81* (21), 4764–4767.
42. Shraiman, B. I. *Proc. Natl. Acad. Sci.* **2005**, *102*, 3318.
43. Elking, J. T.; Reichert, W. M. *Exp. Cell Res.* **2006**, *312*, 2424–2432.
44. Stokes, C. L.; Lauffenburger, D. A.; Williams, S. K. *J. Cell Sci.* **1991**, *99*, 419–430.
45. Subramaniyan, A. K.; Sun, C. T. *Int. J. Solids Struct.* **2008**, *45*, 4340–4346.
46. Zamel, A.; Safran, S. A. *Phys. Rev. E* **2000**, *776*, 0219051-11.
47. Dambo, M.; Vang, Y. *Biophys. J.* **1999**, *76*, 307.
48. Balaban, N. Q.; Schyarz, V. S.; Rineline, D. *Nat. Cell Biol.* **2001**, *3*, 472.

49. Butlu, J.; Tolic-Norrelykke, I. M.; Fabrey, B. *Am. J. Physiol., Cell Physiol.* **2002**, *282*, 595.
50. Yu, Z.; Calvert, T. L.; Leckbant, D. *Biochemistry* **1998**, *37*, 1540.
51. Moora, M.; Cassinelli, C. *J. Biomater. Sci.* **1997**, *9*, 55.
52. Vitte, J.; Benolieli, M.; Pierres, N.; Bongrandi, P. *Eur. Cells Mater.* **2004**, *7*, 52.
53. Coombs, D.; Dembo, M.; Wafs, C. *Biophys. J.* **2004**, *86*, 1408.
54. Chang, G.; Tse, J.; Jain, R. K.; Munn, L. L. *PloS One* **2009**, *4*, e4632.
55. Fray, E. *Adv. Solid State Phys.* **2001**, *41*, 345.

Chapter 15

Interactions between Biopolymers and Polysiloxanes: A Theoretical Study

Eufrozina A. Hoffmann,[1,*] Ljiljana S. Korugic-Karasz,[2] Zoltan A. Fekete,[3] and Tamás Körtvélyesi[1,3]

[1]Department of Physical Chemistry, University of Szeged, Rerrich B. sq. 1., H-6720 Szeged, Hungary
[2]Department of Polymer Science and Engineering, University of Massachusetts, Amherst, MA 01003, USA
[3]HPC group, University of Szeged, Hungary
***efi@chem.u-szeged.hu**

Polysiloxane based devices are frequently used in medical and analytical systems containing biopolymers such as proteins, peptide nucleic acids, polyamides. Although silicones are mainly known for their chemical inertness in living systems, physical adsorption does occur. This is sometimes favorable, for example in gas chromatography, but most of the cases it is a serious obstacle of the proper application. These interactions have been studied by molecular docking.

Introduction

Polysiloxanes are widely used in biological, biomedical and analytical systems e.g. in biosensors, in gas-chromatography (GC), gas-chromatography–mass spectrometry (GC-MS). Controlling biomaterial adsorption to polysiloxanes is an important task of current research interest. However, the theoretical modeling of the interactions between biopolymers and silicones – even on the semiempirical level – is very complex and time consuming, due to the large size of these species. Therefore, in this contribution, we applied as a first step a simplified simulation using molecular dynamics and molecular docking. Molecular docking is a computational simulation of a candidate ligand binding to a receptor/another molecule. It predicts the preferred orientation of one molecule to a second when bound to each other forming a stable complex with association (*1*). The explanation of the enzyme reactions (between the target and the substrate(s)) were explained by the lock-key model (*2*), and its modification the induced fit model (*3*).

Docking has three main approaches (*4–7*). One uses a matching technique that describes the receptor and the ligand as complementary surfaces; the other simulates the actual docking process in which the ligand-protein pair-wise interaction energies are calculated, and the third is the combination of the two previous methods. Typically, the receptor ('host') is a protein or other biopolymer, and the ligand ('guest') is a small molecule (e.g. drug) or another biopolymer. In our models the receptors are different type of polysiloxane chains, and the ligands are selected small segments of biopolymers (tripeptides, trimers of peptide nucleic acid-amino acids and small biogenic polyamides).

Proteins and Polysiloxanes

Because of their inert biologic properties, good mechanical characteristics, transparency and high gas permeability, polysiloxanes (especially poly(dimethylsiloxane) (PDMS)) are widely used in living systems (*8*). They are applied as tissue scaffolds, implants including blood contacting surfaces (*9–13*). Silicone hydrogels are currently used as contact lenses (*14*). In these applications, the ability to prevent the nonspecific adsorption of proteins is essential for proper functioning. However, in some cases the interactions of PDMS with proteins result in their stabilization, which may be useful for drug delivery or screening applications (*15*). The optical and chemical properties of PDMS are specially useful for fabrication of microfluidic devices integrated with optical detection systems (*16–23*), even in 'lab-on-a-chip' systems. But the non-specific protein adsorption is also a serious obstacle delaying its use in heterogeneous immunoassay.

Experimentally the interactions between protein and siloxanes have been studied by Bartzoka *et al.* (*15*) applying Angular-dependent X-ray photoelectron spectroscopy and contact angle measurements; by Bassindale *et al.* (*24*) using of silsesquioxane cages and phage display technology; and by Lok *et al.* (*25*) utilizing total internal reflection fluorescence measurements.

Peptide Nucleic Acids (PNAs) and Polysiloxanes

In the early 1980s Buchardts and Nielsen developed a novel type of compound – peptide nucleic acid (PNA), in which the sugar phosphate backbone is replaced, usually, by N-(2-amino-ethyl)-glycine units with peptide bonds resulting molecules with the property of achirality and hydrophobicity (*26*, *27*). The molecule is an analogoue to DNA. On the side chain, connected to a N atom on the backbone, the natural or modified nucleotide bases are coupled. These compounds are chemically stable and resistant against enzymatic cleavage. Their hybrid complexes with other PNA or DNA chains are stable thermally, and are almost neutral against the change in the ionic strength. They are capable for sequence specific recognition of other DNA or RNA chains with Watson-Crick (WC) hydrogen bonding rules. On the basis of electric discharge experiments, recently it has been proposed that PNAs are the first genetic molecules rather than RNA (*28*, *29*). With changing the structure of backbone, new PNA structures were developed: with the natural peptide backbone structure based on the γ side chain carbon atom, chiral peptide nucleic acids (cPNAs) were synthesized (*30*).

PNA is capable to form homoduplexes, and hybrid complexes with DNA and RNA obeying the WC hydrogen bonding scheme (*31*). These complexes have extraordinary thermal stability and unique ionic strength effects; therefore they can be possibly used as a molecular tool in biotechnology (*32–35*). *In vitro* studies indicate that PNA could inhibit both transcription and translation of genes, thus they are promising candidate for antigen and antisense therapy. For diagnostics purposes they can be used in biosensors (*36–38*).

Because of the possible pharmaceutical and biotechnological applications, it is also an important question how silicon implants interact with PNAs.

Oligoamines and Polysiloxanes

Oligoamines (commonly called polyamines) are organic compounds having two or more amino groups; they play a definitive role in many biological processes such as nucleic acid metabolism, protein synthesis, and cell growth. Spermidine (SPD) is a biogenic polyamine. It seems to promote hair growth, therefore SPD is frequent ingredient in diet integrators and hair lotions. Gambaro *et al.* (*39*, *40*) determined SPD quantitatively applying capillary gas chromatography on methyl silicone capillary column. The samples containing SPD were treated with an alkaline aqueous solution and internal standard was added. The emulsion was extracted with diethyl ether containing ethyl chloroformate, thus the acylated derivative *ethyl[4(ethoxycarbonilamino)butyl]-[3(ethoxycarbonilamino)propyl]carbamate* was formed. Ether extracts, evaporated to dryness and reconstituted in ethyl acetate, were analyzed. Note, Marks and Anderson determinated spermidine in tuna by GC/MS using the same method as well (*41*).

Aims

In this contribution a systematic theoretical investigation has been carried out on the interactions between biopolymers and different polysiloxanes. Our aim was to answer the questions: Can the extent of biopolymer adsorption and/or absorption be predicted? Is it possible to choose chromatographic columns for a given species based on this type of calculations? What are the main interactions leading to the association of biopolymers with polysiloxane? To answer the first question, we are searching for structures of polysiloxanes whose interactions with polypeptides are weak. On the other hand, for the GC separation strong interactions are favorable. Based on our former work (*42*), good correlation has been obtained between McReynolds' constant and the score energy of docking test compounds on methyl- and methyl-phenyl-siloxanes. It was concluded that a lower binding score energy implies better GC separation.

Calculations

To model the polymers of the stationary phases, oligomer chains were constructed by the program PCModel 7.0 (*43*). Each of them contained 16 monomer units. The end groups were trimethylsiloxane (-O-Si(CH_3)$_3$) for the poly(siloxanes). Figure 1 shows the monomer units considered, and Table I. presents the abbreviations and compositions of the modeled silicones. In the first stage of calculations, molecular dynamics simulations were carried out with the program PCModel 7.0/MMX force field with the following parameters: equilibrium time: 1000 fs, run time: 5000 fs, temperature 1500 K (isotherm), heat transfer time 5 fs. These calculations were repeated ten times for every polymer. After each simulation, the energy was minimized by Steepest Descent and Newton-Raphson methods, and the next calculation started from the minimized geometry. The structure with the best energy was used in the docking procedure.

Figures 2.-3. present the structure of ligands; the tripeptides (figure 2a) and PNA fragments (figure 2b.) and SPD and its acylated derivative (figure 3.). The tripeptides were used in protected forms, and contained the following amino acids: alanine (Ac-AAA-NHMe), serine (Ac-SSS-NHMe), phenylalanine (Ac-FFF-NHMe) and tyrosine (Ac-YYY-NHMe). For comparison, we also included an aminoethyl-glycine trimer (the backbone of a single-strand PNA, stripped of its nucleotides), Ac-PNA(A)GG-NHME and Ac-cPNA(A)GG-NHME (see figure 2b). In each cases, the protecting groups were acetyl at the nitrogen and N-methyl amide at the carboxyl terminals.

The structures of the ligands were optimized with the semi empirical method PM6 implemented in the MOPAC2009 program (*44*). AM1-BCC and Gasteiger charges (*45*) were assigned on the atoms of the target and ligand molecules, respectively, without further geometry optimization.

Docking was performed with UCSF DOCK4 (*46*) and DOCK6 (*47*) suites, with random search and random matching (5000 orientations), anchor search and torsion drive. The score energy cutoff distance was chosen essentially infinite (9999 Å), so that no interaction got neglected. All spheres generated were used in the random matching. 2000 iterations with ten cycles were applied to find the best complex. The score function in DOCK is based on a simple 12-6 Van der Waals potential (based on AMBER91 force field and a coulomb function). The target molecules were held rigid, but the ligand molecules were flexible. Both the intra- and intermolecular interactions were considered.

Results and Discussion

Proteins (and PNAs) and Polysiloxanes

The results of tripeptide and PNA fragments docking on silicones are summarized in Table II. The complexes of tripeptide and polysiloxane OV101, which contains only methyl silicone groups, have the highest energy. It means these complexes are the less stable, therefore the smallest absorption is expected in these cases. Figure 4. shows the docked AcO-AAA-NHMe on OV101.

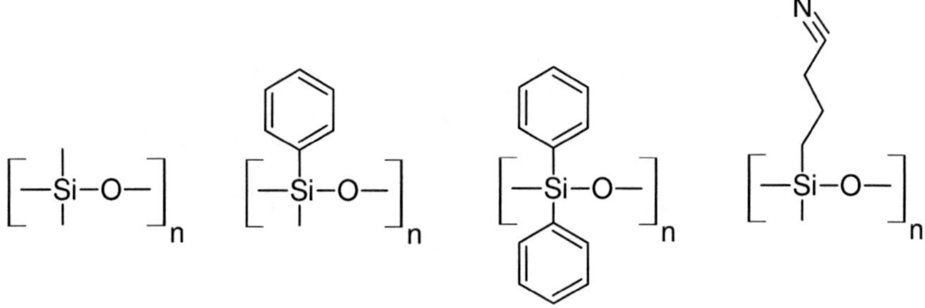

Figure 1. The chemical structures of the applied monomer units

Table I. The compositions of the polysiloxanes

Abbreviation	*methyl silicone*	*phenyl silicone*	*cyanobutyl silicone*
OV101	100%	–	–
OV3	90%	10%	–
OV7	80%	20%	–
OV17	50%	50%	–
OV25	25%	75%	–
OV225	50%	25%	25%

The total score energy of the polymers, consisting of phenyl and methyl silicone as well, are very similar. Generally in the cases of OV7 and OV25 the energies are higher than in the case of OV3 and OV225, although the differences are not too high and the tendency reverses occasionally. The electrostatic energy is typically more negative than the van der Waals energy in these cases; but they have the same order of magnitude.

The silicone OV225 has polar butylciano groups, this is the reason why the electrostatic energy and the total score energy of the complexes with OV225 indicate stronger interactions. Thus if the aim is to prevent protein adsorption then its use is not recommended.

Considering the single strand PNA and polysiloxane interactions, the score energy is a little higher to the tripeptide-silicone complexes. The main difference is that the van der Waals energy clearly dominates here, being 6-40 times more negative than the electrostatic energy. The PNA(A)GG and cPNA(A)GG follow the same pattern. Their interactions are weakest with the non-polar OV101. Since the PNA homoduplexes are dimers of two such chains, it can be expected that similar interactions will be dominating in their case, as well..

Oligoamines and Siloxanes

Table III. shows the results of the docking of spermidine and its derivative on different type of polysiloxanes. It can be seen from the results that the van der

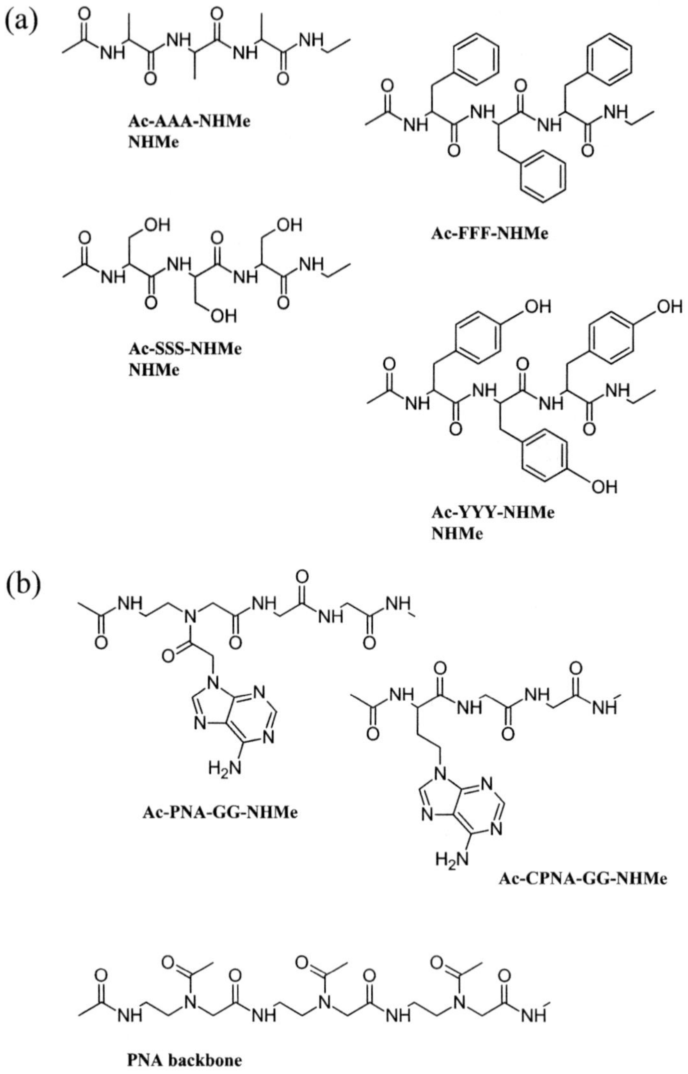

(a)

Ac-AAA-NHMe
NHMe

Ac-FFF-NHMe

Ac-SSS-NHMe
NHMe

Ac-YYY-NHMe
NHMe

(b)

Ac-PNA-GG-NHMe

Ac-CPNA-GG-NHMe

PNA backbone

Figure 2. (a) The docked tripeptides. (b) The PNA fragments.

Waals interactions has the decisive role. The effect of the coulombic interactions appears negligable.

The SPD ligand has a similar behavior with the phenyl-methylsilicones The value of the total score energy is similar. The least negative score energy has been obtained between the SPD (which is an apolar species) and OV225 (which contains polar cyano groups).

We could prove that modifying SPD, in the way described in the introduction, leads to better GC separation, because the scoring energy of the derivate is more negative by 6-9 kcal/mol than the unmodified SPD. The lowest score energy is due to the polymer OV7, which means the gas chromatographic phase containing 20%

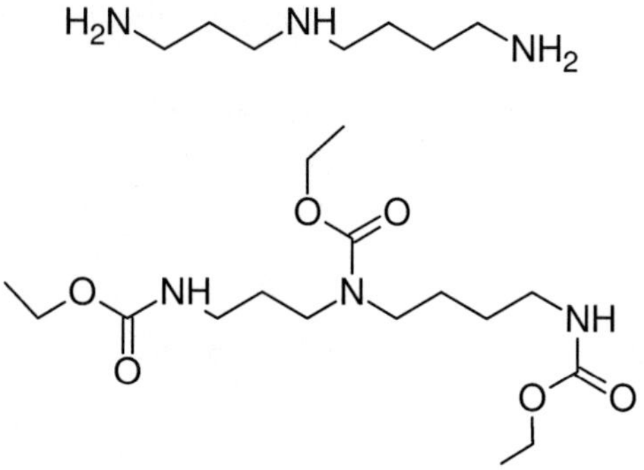

Figure 3. *The chemical structures of spermidine and its derivative*

phenyl- and 80% methyl silicone appears the ideal one for the GC separation of SPD derivative.

Conclusion

We used a theoretical method, molecular docking, to describe interactions between various biopolymers and polysiloxanes. Like every modeling this procedure has its limitations, because of the approximations applied. Only moderate size fragments of polymers was calculated. We used rigid receptors, as generally accepted in the literature. The role of the solvent was neglected, which is correct for modeling GC separations, but it can modify the results of the protein adsorption under analytical or biological conditions. We also disregarded the fact that not only complexes with the lowest energy could form: it is possible to have multiple types of biding site on a receptor. Despite these limitations, the introduced method is general and can give a fast prediction (whose reliability can be improved by refining the approach based upon further data for the specific systems considered). It is a useful tool to support a choice among polysiloxanes for analytical or medical purpose in systems containing biopolymers.

Acknowledgments

This work was supported by the Hungarian Research Fund (OTKA K61577).

References

1. Lengauer, T.; Rarey, M. *Curr. Opin. Struct. Biol.* **1996**, *6* (3), 402.
2. Fisher, E. *Ber. Dtsch. Chem. Ges.* **1894**, *27*, 2985.
3. Koshland, D. E. *Proc. Natl. Acad. Sci.* **1958**, *44* (2), 98.
4. Lang, P. T.; Brozell, S. R.; Mukherjee, S.; Pettersen, E. T.; Meng, E. C.; Thomas, V.; Rizzo, R. C. Case, D. A.; James, T. L.; Kuntz, I. D. *RNA*; **2009**, DOI: 10.1261/rna.1563609.
5. Moustakas, D.; Lang, P. T.; Pegg, S.; Pettersen, E. T.; Kuntz, I. D.; Broojimans, N.; Rizzo, R. C. *Proc. Natl. Acad. Sci.* **1958**, *44* (2), 98.

6. Meng, E. C.; Shoichet, B. K.; Kuntz, I. D. *J. Comput. Chem.* **2004**, *13* (4), 505.
7. Ewing, T. J. A.; Kuntz, I. D. *J. Comput. Chem.* **1997**, *18*, 1175.
8. Chena, H.; Brooka, M. A.; Sheardownb, H. *Biomaterials* **2004**, *25*, 2273.
9. Childs, M. A.; Matlock, D. D.; Dorgan, J. R.; Ohno, T. R. *Biomacromolecules* **2001**, *2*, 526.
10. Belanger, M. C.; Marois, Y. *J. Biomed. Mater. Res.* **2001**, *58*, 467.
11. Chen, K. Y.; Kuo, J. F.; Chen, C. Y. *J. Biomater. Sci., Polym. Ed.* **1999**, *10*, 1183.
12. Chen, Z.; Ward, R.; Tian, Y.; Malizia, F.; Gracias, D. H.; Shen, Y. R.; Somorjai, G. A. *J. Biomed. Mater. Res.* **2002**, *62*, 254.
13. Park, J. H.; Bae, Y. H. *Biomaterials* **2002**, *23*, 1797.
14. Kunzler, J. F. *Trends Polym. Sci.* **1996**, *4*, 52.
15. Bartzoka, V.; Chan, G.; Brook, M. A. *Langmuir* **2000**, *16*, 4589.
16. Eteshola, E.; Leckband, D. *Sens. Actuators, B* **2001**, *72*, 129.
17. Philips, K. S.; Cheng, Q. *Anal. Chem.* **2005**, *77*, 327.
18. Quake, S. R.; Scherer, A. *Science* **2000**, *290*, 1536.
19. Fujii, T. *Microelectron. Eng.* **2002**, *61–62*, 907.
20. Unger, M. A.; Chou, H. P.; Thorsen, T.; Scherer, A.; Quake, S. R. *Science* **2000**, *288*, 113.
21. McDonald, J. C.; Whitesides, G. M. *Acc. Chem. Res.* **2002**, *35*, 491.
22. Thorslund, S.; Sanchez, J.; Larson, R.; Nikolajeff, F.; Bergquist, J. *Colloids Surf., B* **2005**, *46*, 240.
23. Berdichevsky, Y.; Khandurina, J.; Guttman, A.; Lo, Y. H. *Sens. Actuators, B* **2004**, *97*, 402.
24. Bassindale, A. R.; Codina-Barrios, A.; Frascione, N.; Taylor, P. G. *New J. Chem.* **2008**, *32*, 240.
25. Lok, B. K.; Cheng, Y.; Robertson, C. R. *J. Colloid Interface Sci.* **1983**, *91* (1), 104.
26. Nielsen, P. E.; Egholm, M.; Berg, R. H.; Buchardt, O. *Science* **1991**, *254*, 1497.
27. Egholm, M.; Buchardt, O.; Nielsen, P. E.; Berg, R. H. *J. Am. Chem. Soc.* **1992**, *114*, 1895.
28. Nelson, K. E.; Levy, M.; Miller, S. L. *Proc. Natl. Acad. Sci.* **2000**, *97* (8), 3868.
29. Srivatsan, S. G. *Pure Appl. Chem.* **2004**, *76* (12), 2085.
30. Lenzi, A.; Reginato, G.; Taddei, M.; Trifilieff, E. *Tetrahedron Lett.* **1995**, *36*, 1717.
31. Egholm, M.; Buchardt, O.; Christensen, L.; Behrens, C.; Frier, S. M.; Driver, D. A.; Berg, R. H.; Kim, S. K.; Nordén, B.; Nielsen, P. E. *Nature* **1993**, *365*, 566.
32. Demidov, V.; Frank-Kamenetskii, M. D.; Egholm, M.; Buchardt, O.; Nielsen, P. E. *Nucleic Acids Res.* **1993**, *21*, 2103.
33. Demers, D. B.; Curry, E. T.; Egholm, M; Sozer, A. C. *Nucleic Acids Res.* **1995**, *23*, 3050.
34. Örum, H.; Nielsen, P. E.; Jorgensen, M.; Larsson, C.; Stanley, C.; Koch, T. *BioTechniques* **1995**, *19*, 472.

35. Veselkov, A. G.; Demidov, V.; Nielsen, P. E.; Frank-Kamenetskii, M. D. *Nucleic Acids Res.* **1996**, *24*, 2483.
36. Carlsson, C.; Jonsson, M.; Nordén, B.; Dulay, M. T.; Zare, R. N.; Noolandi, J.; Nielsen, P. E.; Tsui, L-C.; Zielenski, J. *Nature* **1996**, *380*, 207.
37. Thiede, C.; Bayerdörffer, E.; Blasczyk, R.; Wittig, B.; Neubauer, A. *Nucleic Acids Res.* **1996**, *24*, 983.
38. Wang, J.; Palecek, E.; Nielsen, P. E.; Rivas, G.; Cai, X.; Shirashi, H.; Dontha, N.; Luo, D.; Farias, P. A. M. *J. Am. Chem. Soc.* **1996**, *118*, 7667.
39. Gambaro, V.; Casagni, E.; Dell'Acqua, L.; Valent, M.; Visconti, G. L. *J. Pharm. Biomed. Anal.* **2004**, *35* (2), 409.
40. Gambaro, V.; Casagni, E.; Dell'Acqua, L.; Valent, M.; Visconti, G. L. *J. Sep. Sci.* **2006**, *29* (9), 1294.
41. Marks, H. S.; Anderson, C. R. *J. AOAC Int.* **2006**, *89* (6), 1591.
42. Körtvélyesi, T.; Pálinkó, I.; Hoffmann, E. A. Manuscript prepared for publication.
43. *PCModel*, version 7.0; Serena Software: Bloomington, IN, 2007.
44. Stewart, J. J. P. *J. Mol. Model.* **2007**, *13*, 1173.
45. *Ambertools*, version 1.2; University of California: San Francisco, 2006.
46. *DOCK Suite*, version 4.0.1; University of California: San Francisco, 1998.
47. *DOCK Suite*, version 6.2; University of California: San Francisco, 2008.

Table II. The docking energy of tripeptides on different type of polysiloxanes[a]

		OV101	*OV3*	*OV7*	*OV17*	*OV25*	*OV225*
Ac-AAA-NHMe	Score	**-22.30**	**-37.70**	**-30.22**	**-37.35**	**-31.01**	**-57.29**
	vdW	-16.37	-7.13	-14.82	-12.91	-14.23	-12.00
	ES	-5.93	-30.57	-15.40	-24.44	-16.78	-45.29
Ac-SSS-NHMe	Score	**-21.68**	**-37.43**	**-30.20**	**-36.57**	**-31.44**	**-53.6**
	vdW	-13.21	-12.21	-12.53	-13.3	-15.25	-15.28
	ES	-8.46	-25.23	-17.67	-23.24	-16.19	-38.36
Ac-FFF-NHMe	Score	**-27.92**	**-38.87**	**-33.79**	**-37.82**	**-33.95**	**-57.74**
	vdW	-20.38	-13.42	-18.8	-15.16	-16.47	-13.66
	ES	-7.54	-25.45	-15.5	-22.66	-17.48	-44.08
Ac-YYY-NHMe	Score	**-29.23**	**-37.78**	**-36.28**	**-33.79**	**-34.49**	**-58.54**
	vdW	-20.40	-13.64	-20.11	-18.20	-18.00	-18.46
	ES	-8.83	-24.14	-16.17	-15.59	-16.49	-40.08
Single-strand PNA	Score	**-25.41**	**-31.8**	**-30.78**	**-32.53**	**-30.13**	**-26.91**
	vdW	-24.62	-30.17	-30.05	-26.57	-25.89	-23.78
	ES	-0.79	-1.69	-0.73	-5.96	-4.23	-3.13
Ac-PNA(A)GG-NHME	Score	**-24.31**	**-32.00**	**-33.39**	**-26.24**	**-23.57**	**-32.00**
	vdW	-24.30	-31.36	-31.06	-24.30	-19.93	-25.12
	ES	-0.01	-0.63	-2.32	-1.94	-3.64	-6.88
Ac-cPNA(A)GG-NHME	Score	**-23.85**	**-28.12**	**-31.45**	**-26.98**	**-25.83**	**-24.05**
	vdW	-22.57	-27.28	-28.86	-24.11	-22.78	-19.87
	ES	-1.28	-0.84	-2.59	-2.86	-3.06	-4.18

[a] Units are kcal/mol. Score denotes the total score energy, vdW is the abbreviation of the van der Waals energy, and ES indicates the electrostatic (Coulomb) energy. For the compositions of OV phases see Table I.

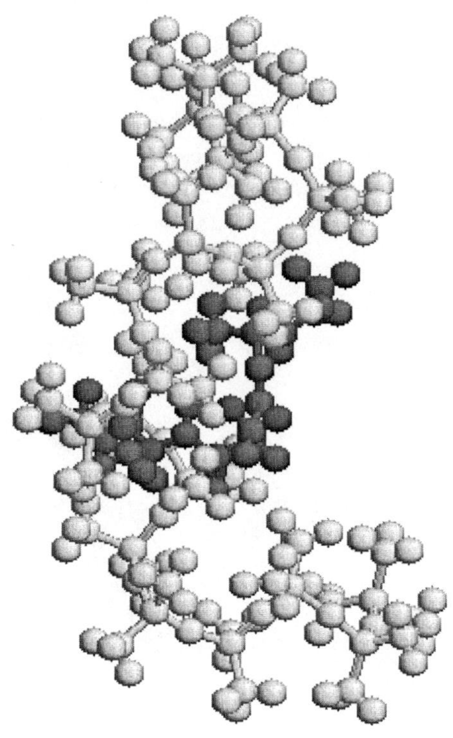

Figure 4. The docked AcO-AAA-NHMe (dark) on OV101 (light grey)

Table III. The docking energy of spermidine and its derivative on different type of polysiloxanes[a]

Abbr. of silicones	*Spermidine*			*spermidine derivative*		
	Score	*vdW*	*ES*	*Score*	*vdW*	*ES*
OV101	-12.28	-12.42	0.14	-21.46	-20.51	-0.95
OV3	-14.67	-13.61	-1.06	-20.37	-19.29	-1.08
OV7	-13.74	-13.01	-0.73	-22.73	-22.52	-0.21
OV17	-13.34	-12.97	-0.36	-21.14	-20.06	-1.08
OV25	-13.70	-12.32	-1.37	-14.16	-14.30	0.14
OV225	-9.74	-9.18	-0.56	-18.10	-17.35	-0.75

[a] Units are kcal/mol. Score denotes the total score energy, vdW is the abbreviation of the van der Waals energy, and ES indicates the electrostatic (Coulomb) energy. For the compositions of OV phases see Table I.

Indexes

Author Index

Subject Index

C

D

I

L

M

N

O

P

R

S